DISCARDED

TOWARDS A PHILOSOPHY OF CRITICAL MATHEMATICS EDUCATION

Mathematics Education Library

VOLUME 15

Managing Editor

A.J. Bishop, *Monash University, Melbourne, Australia*

Editorial Board

H. Bauersfeld, *Bielefeld, Germany*
J. Kilpatrick, *Athens, U.S.A.*
C. Laborde, *Grenoble, France*
G. Leder, *Melbourne, Australia*
S. Turnau, *Krakow, Poland*

TOWARDS A PHILOSOPHY OF CRITICAL MATHEMATICS EDUCATION

by

OLE SKOVSMOSE
Aalborg University, Denmark

KLUWER ACADEMIC PUBLISHERS
DORDRECHT / BOSTON / LONDON

A C.I.P. Catalogue record for this book is available from the Library of Congress.

ISBN 0-7923-2932-5

Published by Kluwer Academic Publishers,
P.O. Box 17, 3300 AA Dordrecht, The Netherlands.

Kluwer Academic Publishers incorporates
the publishing programmes of
D. Reidel, Martinus Nijhoff, Dr W. Junk and MTP Press.

Sold and distributed in the U.S.A. and Canada
by Kluwer Academic Publishers,
101 Philip Drive, Norwell, MA 02061, U.S.A.

In all other countries, sold and distributed
by Kluwer Academic Publishers Group,
P.O. Box 322, 3300 AH Dordrecht, The Netherlands.

Printed on acid-free paper

All Rights Reserved
© 1994 Kluwer Academic Publishers
No part of the material protected by this copyright notice may be reproduced or
utilized in any form or by any means, electronic or mechanical,
including photocopying, recording or by any information storage and
retrieval system, without written permission from the copyright owner.

Printed in the Netherlands

CONTENTS

Acknowledgements ix

Introduction 1

Chapter 1: Critique and Education 11
1. Crisis 12
2. Critique 14
3. Emancipation? 19
4. Critical Education 21
5. Literacy and Mathemacy 24

Chapter 2: Democracy and Education 28
1. Links between 'Democracy' and 'Education' 28
2. Basic Democratic Ideas 31
3. Democratic Competence 34
4. A Problem of Democracy in a Highly Technological Society 38
5. Education for *Mündigkeit* 40

Chapter 3: Mathematics – A Formatting Power? 42
1. Technology and the Vico Paradox 43
2. Mathematics and Technology 48
3. Abstractions 50
4. Formalisations 53
5. Mathematics as Critical 56

Chapter 4: A Thematic Approach in Mathematics Education 59
1. Some Sources of Inspiration 60
2. Planning a Thematic Approach 62
3. Comments on the Project 68
4. The Diary and the 'Results' 70
5. Exemplarity 73

Chapter 5: "Golfparken" and "Constructions"   79

1. Opinions about Mathematics   80
2. "Golfparken"   82
3. "Construction"   86
4. Comments on the Projects   90

Chapter 6: Reflective Knowing   97

1. Reflective Knowing: A First Delineation   98
2. Reflections and Modelling   102
3. Reflective Knowing in Educational Practice   114
4. Six Entry Points to Reflective Knowing   118
5. A Note about 'Knowing'   122

Chapter 7: "Family Support in a Micro-Society"   125

1. The Structure of the Project   125
2. Comments on the Project   129
3. Reflective Knowing in the Project   133
4. Understanding 'Formatting'   136
5. A Note about Challenging Questions   138

Chapter 8: "Our Community"   141

1. The Structure of the Project   142
2. Comments on the Project   148
3. Reflective Knowing – An Open Concept   152

Chapter 9: "Energy"   155

1. The Structure of the Project   155
2. Comments on the Project   163
3. Formal Language versus Natural Language   166
4. Comments on Mathemacy   170

Chapter 10: Intentionality   175

1. Dispositions, Intentions and Actions   176
2. Learning as Action   181
3. Different Forms of Epistemic Development   184
4. Personal Fatalism, Servility and Achievement   189
5. Some Comments on the Projects   192

CONTENTS

Chapter 11: Knowing, An Epilogue	196
1. Knowledge – A Controlled Concept	196
2. Knowing – An Open Concept	199
3. Mono-Logical Epistemic Theories	201
4. Dia-Logical Epistemic Theories	205
5. Knowing – An Explosive Concept	206
Notes	210
Bibliography	227
Name Index	237
Subject Index	241

ACKNOWLEDGEMENTS

In 1988 the Danish Research Council for the Humanities decided to set up the research initiative 'Mathematics Education and Democracy in Highly Technological Societies'. The overall intention of this project was to discuss mathematics education as part of democratic endeavour in a highly technological society. The initiative ran for a period of five years. The members of the group directing the initiative were Gunhild Nissen (chairperson), University of Roskilde; Jens Bjørneboe (secretary); Morten Blomhøj (secretary), University of Roskilde; Peter Bollerslev, Ministry of Education; Vagn Lundsgaard Hansen, Technical University of Copenhagen; Mogens Niss, University of Roskilde; and Ebbe Thue Poulsen, University of Aarhus. My work forms a part of this overall initiative. It has been supported financially by the Danish Research Council and the University of Aalborg; additional support has been received from Den Obelske Familiefond.

My attempt to develop a philosophy of a critical mathematics education has had much help and inspiration from experimental work carried out in relation to the overall initiative. The teachers and schools who have contributed the most to my work include Henning Bødtkjer, Klarup skole; Jens Jørgen Andersen, Ole Dyhr and Thue Ørberg, Aalborg Friskole; Ib Trankjær, Nyvangskolen in Randers; and Jørgen Boll and Jørgen Vognsen, Rønbækskolen in Hinnerup. In addition Andreas Reinholt from Aalborg Teacher Training College has co-ordinated much of the experimental work. Several researchers have been involved in 'Mathematics Education and Democracy in Highly Technological Societies'. In particular, I have had inspiration and support from Morten Blomhøj and Lena Lindenskov, University of Roskilde; Dan Eriksen, Royal Danish School of Educational Studies; Kirsten Grønbæk Hansen, University of Copenhagen; and Helle Alrø and Iben Maj Christiansen, University of Aalborg. Karin Beyer, University of Roskilde, has given me useful suggestions for improving parts of the manuscript. In addition I have received most valuable criticism, of the whole manuscript, from Marcelo Borba, State University of Sao Paulo at Rio Claro.

I want to thank the directors of the initiative as well as the teachers and the researchers involved for their support as well as for their comments on both my work and my manuscript.

During the year 1990–1991, I visited the Department of Education, University of Cambridge where I gained much advice from Alan Bishop, not only about how to begin my investigation and how to finish the book, but also about what comes in between. Alan Bishop has made it possible for me to publish my work in the Mathematics Education Library. Marilyn Nickson, University of Cambridge, has carefully gone through the whole manuscript and made the final linguistic corrections. I want to express my gratitude to both Alan Bishop and Marilyn Nickson for their encouragement and efforts.

Aalborg, November 1993
Ole Skovsmose

INTRODUCTION

In *Nineteen Eighty-Four* George Orwell gives a description of different forms of suppression. We learn about the telescreens placed everywhere, through which it is possible for Big-Brother to watch the inhabitants of Oceania. However, it is not only important to control the activities of the inhabitants, it is important as well to control their thoughts, and the Thought Police are on guard. This is a very direct form of monitoring and control, but Orwell also outlines a more imperceptible and calculated line of thought control.

In the Appendix to *Nineteen Eighty-Four* Orwell explains some structures of 'Newspeak', which is going to become the official language of Oceania. Newspeak is being developed by the Ministry of Truth, and this language has to substitute 'Oldspeak' (similar to standard English). Newspeak should fit with the official politics of Oceania ruled by the Ingsoc party: "The purpose of Newspeak was not only to provide a medium of expression for the world-view and mental habits proper to the devotees of Ingsoc, but to make all other modes of thought impossible. It was intended that when Newspeak had been adopted once and for all and Oldspeak forgotten, a heretical thought – that is, a thought diverging from the principles of Ingsoc – should be literally unthinkable, at least as far as thought is dependent on words. Its vocabulary was so constructed as to give exact and often very subtle expression to every meaning that a Party member could properly wish to express, while excluding all other meanings and also the possibility of arriving at them by indirect methods. This was done partly by the invention of new words, but chiefly by eliminating undesirable words and by stripping such words as remained of unorthodox meanings, and as far as possible of all secondary meanings whatever. To give a single example. The word *free* still existed in Newspeak, but it could only be used in such statements as 'This dog is free from lice' or 'This field is free from weeds'. It could not be used in its old sense of 'politically free' or 'intellectually free', since political and intellectual freedom no longer existed even as concepts, and were therefore of necessity nameless."[1]

I am not going to consider life in Orwell's Oceania, nor the activity of the Thought Police or the role of Big-Brother; but I want to take a closer look at the presumption that a language could prestructure a world-view and exclude some modes of thought. Does it make sense to think of language not only as a flexible medium for expressing ideas but also as a filter for formulating ideas? Does it make sense to try to control thoughts by restructuring a language? Does the structure of a language specify the nature of thoughts which can be formulated in the language?

In the *Tractatus Logico-Philosophicus* by Ludwig Wittgenstein we meet another thesis about the influence of language. Wittgenstein found that ordinary language contains ambiguous notions, and that this phenomenon is the main cause of many philosophical problems. Such problems in fact take the shape of linguistic misunderstandings. To prevent such confusions, it is useful to introduce a formal language, and Wittgenstein refers to the symbolism developed by Gottlob Frege and Bertrand Russell.[2] The introduction of such a formalism excludes the possibility for linguistic misunderstandings, and therefore classical philosophical problems will be excluded as well. In this way a formal language comes to exclude some modes of thought. In the *Tractatus* Wittgenstein also states that the limits of our language are the limits of our world.[3]

The idea that a formalised language may be useful in excluding certain formulations and supposed misunderstandings is developed further by logical positivism. If logical positivism had been able to fulfil its programme, the language of science would contain exactly those statements which have a meaning according to the principle of verification, i.e. all statements which can be checked against empirical evidence.[4] According to some trends in logical positivism (not all logical positivists agreed with this) normative sentences do not constitute proper assertions at all, and consequently ethics is to be excommunicated from the language of sciences.

These considerations relate to the idea of *linguistic relativism*: the language available determines the thought we have the potential to express. A classical interpretation of this relativism is found in the works of Edward Sapir and Benjamin Lee Whorf. Sapir writes: "Human beings do not live in the objective world alone, nor alone in the world of social activity as ordinarily understood, but are very much at the mercy of the particular language which has become the medium of expression for their society. It is quite an illusion to imagine that one adjusts to reality essentially without the use of language and the language is merely an incidental means of solving specific problems of communication or reflection. The fact of the matter is that the 'real world' is to a large extent unconsciously built up on the language habits of the group."[5] This formulation emphasises the force of introducing Newspeak. If language is more than a means for communication and the 'real world', to a large extent, is built up by the language habits of the group, then to transform language becomes a forceful act. Not only is it true that what we express about the world depends on our language, but the world itself does as well. Whorf emphasises that language is not merely a reproduction instrument for 'voicing ideas' but is rather itself the shaper of ideas. Formulations of ideas are not independent processes, but become part of a particular grammar.[6] Therefore, the users of markedly different grammars are directed, by their grammars, towards different types of observations and towards different evaluations of externally similar acts of observation. They are not equivalent to each other as observers because they advance different views of the world.[7]

It might be the case that some particular observations cannot be made because the language available does not allow the observer to describe some phenomena. Some concepts may be lacking. For instance, the language spoken in Greenland has many more words for ice than English or Danish, which makes it possible for the people there not only to make more accurate descriptions, but (and this is the essential point) also to make more specific observations than a person settled in an English or a Danish language community could make. The possibility of expressing observations is not all that is determined by language, but so are the observations themselves. However, as emphasised by Sapir and Whorf, the relativism of language is connected not only to the available sets of words but to the grammar as well. This makes a more fundamental relativism possible: A grammar influences the possibilities available for what we can express and the purposes for which we can use our language. At the most general level, the thesis of relativism approximates the idea of Friedrich Nietzsche that language may be a dumping ground for all sorts of misconceptions, illusions, fallacies, myths and errors. We have to distrust the whole of language. Not only the misconceptions of our own generation, but misconceptions from previous generations are sedimented into layers of prejudice and ideology. And because language is the medium of thought, our possibilities for grasping the essence of life and existence are poor. We are at the mercy of our mischievous language – and the thesis of linguistic relativism expands into the thesis of epistemic relativism.

The formulations of linguistic relativism can be rephrased in Kantian terms by giving Immanuel Kant a 'linguistic turn'. Kant tries to identify some uniform and *a priori* conditions for experience and for obtaining knowledge. In the *Critique of Pure Reason* he describes some basic forms of intuition and some categories into which all our thinking must fit. It is not just an empirical fact that we perceive physical objects in a space of three dimensions. Space is not a concept generalised from the sum of our experiences. It is a necessary condition for experiences. Our experiences are moulded by *a priori* forms of intuition, as water will take the shape of the pitcher into which it is poured. That means that *a priori* structures organise what is possible for humanity to observe. The fingerprints of humanity, as embodied by the *a priori* structures, are already imposed on what we observe. However, Kant's perspective may be altered: structures of the eternal and ideal Kantian profile may be substituted by structures belonging to our social reality. The *a priori* filters may be substituted by *a posteriori* ones. Some social construction may be interposed between our observations and the 'outer world', the language into which we are submerged. Thus the structure of our language becomes a socially constructed, prestructuring principle of our observations.

Mathematics may be characterised in different ways – including as a language. As such it becomes an instrument for knowledge development and an interpreter of social reality. Therefore, it makes sense to ask whether mathematics can be seen as a language with a universal descriptive power, or if it creates and expands blind spots. How would the world look for a person whose mother tongue is mathematics? A strong belief in the omnipotence of mathematics is expressed by the Galilean formulation that God has organised the world in accordance with the principles of mathematics. If this is true, mathematics must then possess an unlimited and complete descriptive power. However, if God did not organise the world in accordance with mathematics alone, it would make sense to ask if the language of mathematics provides a fruitful or a restricted world view. Does mathematics keep our imagination on a short leash?

Does mathematics make us see reality in a distorted way? If so, distorted in relation to what? Does it make sense to state, as Sapir did, that the 'real world' to a large extent, unconsciously, is built up by the (mathematical) language habits of the group? Could mathematics be interpreted as a language by means of which we not only observe certain structures of reality and ignore others, but also organise reality? And who are the 'we' in this instance? All mathematicians? Or could it be the mathematicians at all, whose 'job' may simply be to develop the grammar of the mathematical language? Could the 'we' be just the applied mathematicians, the technologists and the engineers who are the users of mathematics? Or could the 'we' be given an even more general interpretation? The thesis of linguistic relativism has been formulated in relation to socially defined groups of persons, but if mathematics is a language of science, what then would be the interpretation of linguistic relativism? Does it make sense to say that a science restructures the world?

If we accept that a language need not be the mother tongue of a certain group of people to be interpreted in accordance with the thesis of linguistic relativism, we may look at mathematics as a language with (potentially) a strong social power. Mathematics is connected to our technological culture, and mathematical research has grown enormously and in parallel with the scientific technological revolution. It is, however, far from obvious how the connection between mathematics and technology can be spelled out. Nevertheless, the importance of mathematics, although not comprehended in all its detail, is generally accepted far outside the mathematical and technological community. Students[8] meet mathematics in the curriculum, knowing that mathematical skills are important for their future career. Politicians know that mathematics is one of the pillars supporting technological development, so they accept having to spend great sums of money on this science, which is at the same time held to be the most abstract of all sciences.[9] Mathematics is generally recognised as important, although it is most difficult to point out which specific elements of the technological society this science supports.

Therefore, it becomes important to look at mathematics education because this part of education provides an introduction to a language held as being omnipotent. Does this mean that mathematics education becomes important not only as an introduction to the next generations of some basic stock of knowledge, but also as an introduction to a certain grammar which promotes some specific world views? Does this mean that the importance of mathematics education in our technological society has to be interpreted along the same lines as would a general teaching of Newspeak in Oceania, as making an introduction to some mode of thought? Orwell describes Newspeak as an obstacle to critical thought, and it is possible that the learning of the grammar of mathematics should be approached in the same way. Does the teaching of the powerful mathematical language prevent a critical interpretation of a highly technological society?

It must be emphasised that I do not suggest any comparison of living in a highly technological society with living in Oceania. It would be too simple to transform the antagonism towards Oceania into an aversion to our highly technological society. I am not analysing the question of whether or not social and technological development take us in the direction of a Utopia or closer to Orwell's Dystopia. We shall not look for a Utopia in the past as a romantic union of humanity and nature not spoiled by technological inventions, nor in the future as the complete technical solution to present day problems. None of the variants of this sort of romanticism is part of my project. I simply want to emphasise the importance of making the broadest interpretation of what may be going on in mathematics education, because mathematics education also entails an introduction to a mode of speech and thought and provides an introduction to a certain culture. To understand the significance of a Newspeak education in Oceania, we have to discuss the functions of that language in a society. It is not sufficient to come up with suggestions for improving that particular kind of education. The same is the case for mathematics education in a highly technological society, even if we were to make an optimistic interpretation of the present trends of technology. Mathematics education can be interpreted as part of a symbolic struggle, and it is important to note that a condition for understanding what is going on in that sort of education is not to believe one's own eyes. What seems to happen does not happen, and what happens is difficult to see.

Mathematics education is paid an enormous amount of attention by all institutions in society. It is taken up globally – in the highly technological societies with reference to the importance of keeping pace with social development, and in the developing countries with reference to the demands of making technological progress. Today nobody seems to doubt the general importance of mathematics education. It is important therefore to look at mathematics education in its most general and global perspective.

In the life of universities a demarcation between sciences and the humanities has been observed, for instance by C. P. Snow in *The Two Cultures: And a Second Look*. Snow finds the demarcation line runs across even the high tables in the dining halls at the enchanting colleges in Cambridge. The scientist and the humanist do not communicate because they cannot communicate. They belong to two separate cultures – only frozen smiles are passed across the border. I find it important to take a look at this line of demarcation which exists in our highly technological society, and also to break it down, if we wish to be more than onlookers of technological growth. One area in which something can be done is that of mathematics education. This ought to be of interest to both camps. But mathematics education must be investigated not just as a relationship between a certain subject matter, mathematics, and the students who have to swallow prefabricated bits of that subject. The perspective has to be broader, and the content of mathematics education must be interpreted in such a way as to understand the global role of mathematics. Therefore, we have to overcome this demarcation line.

I shall try to develop elements of a philosophy of critical mathematics education. This needs some explanation. What could be the meaning of 'a philosophy of'? A philosophy does not rest upon facts in any straightforward way. It is a simple mistake to try to improve a philosophical investigation by applying more accurate instruments. It is not possible to support a philosophical thesis by extending empirical evidence. Is it at all possible to give that sort of evidence to philosophical theses? And what about the other way around: Could philosophy provide evidence for statements with an empirical content? Could philosophy support a non-philosophical statement – a statement about educational methodologies, for instance? (Is it possible to refute a philosophy by empirical means?)

What then can we expect a philosophy to do? I do not find it capable of proving very much. By rationalism, as expressed for instance by Baruch de Spinoza, it was assumed that it was possible to prove even the existence of God. But even if that type of ambitious claim is modified, we sometimes meet the supposition that some general principle (for instance concerning ethical values and norms) can be discussed by philosophy, and that it is possible to establish some degree of philosophical justification of such a principle. I am not sure about this possibility. In particular, I do not think that a philosophy can identify educational principles and some justification of these principles. I even doubt the existence of genuine philosophical theses (with a few exceptions perhaps). However, a philosophical investigation may provide some foundations for making interpretations and clarifications (a philosophical thesis?). I look at philosophy as a means of providing new perspectives. It may raise different views by establishing new baselines in our language.

However, even with that unpretentious belief in the potential of philosophy, I do not expect philosophy to be independent of experiences. If we accept a relativism of language – which by itself is an interpretation (supported by observations or by philosophical arguments?) – we may expect interpretations to affect our observations, and what this means is that reflections on the (always interpreted) empirical facts can indicate a criticism of the interpretations used. If some discrepancies appear between (interpreted) facts and theory, they may point towards new schemes of interpretations, and this means a new philosophical development. I see the real job of philosophy as being to bend the bars of the cage into which we are locked by language. But because we shall never be able to escape language, philosophy becomes the task of a Sisyphus.

In mathematics education, seen as a scientific enterprise, there seems to be a traffic jam in the streets connecting theory and practice. At the loading docks of philosophy, lorries arrive loaded with support for some educational reform or stuffed with fully ripe guidelines for curriculum development. From another direction, we find cars returning loaded with experiences telling us that some general educational and philosophical principles are questionable. Others return with supporters maintaining that a variety of observations endorse some overall theses about learning and schooling. Not much seems to regulate this heavy traffic. Nevertheless, I do not find this traffic to be without some usefulness, although it is chaotic, but I think that much of the traffic jam arises because the importance of philosophy as a guideline for reform is overestimated. Philosophy may provide clarification and supply new interpretations, but basic guidelines for educational reforms must be produced by educational practice. However – and this is important – it must be a reflective educational practice. Philosophy can help us to see what may be hidden behind what is immediately visible. What seems to happen does not happen: Do not assume that mathematics education primarily has to do with teaching students mathematics! Philosophy may provide new interpretations of a practice, and a practice may provide a richer meaning to philosophical terms.

I do not find it possible to build up any hierarchical educational theory crowned with a certain philosophy expressing some paramount values. It is not possible to set up some thesis and then to step aside and argue, by purely philosophical means, that this same thesis is sound. I do not think a philosophy of mathematics education can give advice about what to do in the classroom. Again, I have to modify. A philosophy of mathematics education cannot exist in a reclusive way. If philosophy has to do with interpretation and also with the interpretation of observations, then philosophy is 'inside' and becomes part of our observations. But the *raison d'être* of a philosophy of mathematics education is not to provide guidelines, it is to create new possibilities for interpretation, i.e. to create an improved understanding of what is going on in mathematics education. Let this delineate what I intend to be the meaning of 'a philosophy of mathematics education'.

I shall try to establish an 'equality' between educational examples and the development of a philosophical analysis. The analytical relationship between a philosophical clarification and the forms of practice related to it is not that of master and pupil. This means the reader should not try to look for the argument 'because such and such lies within in the philosophy of mathematics education, then a certain form of educational practice is desirable'. I do not think in terms of 'implication' but in terms of 'illumination'. Philosophy may illuminate a practice, and a practice may make us see new things in a philosophy. The examples described in this book have to be seen as ways of providing concepts of philosophical investigations with (educational) meaning. They belong to the semantics of the language of philosophy. Trends in analytical philosophy have sometimes ignored this essential point: Where do the meanings of concepts used in philosophical investigations come from?

The method of providing meaning in philosophical investigations by paying attention to examples has, for instance, been developed by Wittgenstein in his *Philosophical Investigations* and in the lectures which he gave in Cambridge after his 'second coming'. The method of clarifying by means of examples became characteristic of his philosophical analyses. It became not only a peculiar style but also an important means for allowing the analyses of language games to become fraught with meanings. I sympathise with this strategy, but it does have a shortcoming. Wittgenstein has been criticised for not being able to do anything outside philosophy by means of his philosophy. In fact it is possible to observe fundamental similarities between his outlook in the *Tractatus* and the *Investigations*, although it has been emphasised that a gulf separates the two works: While the *Tractatus* talks about 'the language' and makes a fetish of formal language, the *Investigations* talks about 'languages' and provides a relativistic attitude; and while the *Tractatus* develops a 'picture-theory' of language and relates meaning to reference, the *Investigations* describes the meaning of a sentence in terms of its use. Nevertheless, both works suggest that the job of philosophy is to clear up uses and misuses of language. Philosophical problems are caused by misuses of language and philosophical activity consists in identifying such misuse. And – using the perspective of the *Investigations* – a good strategy for identifying types of misuse is the invention of examples of different (simple and artificial) language games.

I do not find that the primary reason Wittgenstein comes to rest in the conservative camp of analytical philosophy, not trying to struggle with the (political) world, is that he refuses to come up with certain philosophically grounded dogmas, although this has been assumed in much criticism of analytical philosophy. I do not think that philosophy can be radicalised by making it provocative and dogmatic. My hypothesis is that one reason for the conservatism of much analytical philosophy is not its lack of dogma, but its lack of serious examples. In Wittgenstein's version they come to belong to 'pretend' realities. I find it important to relate a philosophical discussion to serious examples. Then philosophy not only comes to attack linguistic misunderstandings but also the basic limitations produced by our schemas of interpretation. (And how do I support this philosophical point about philosophy?) Language is a medium in which we are submerged, but I do not subscribe to the 'myth of the framework'. We are not locked into our preconceptions. Even the most basic schemas of interpretations can be challenged.

I have tried to select examples for this book which are interesting from the perspective of critical mathematics education, not that they need always to be 'exemplary' ones, but in the sense that they may help to expose a meaning of such an education. Wittgenstein claims that his small, and often droll, examples have to be analysed seriously. I agree, but I find that examples also have to express the serious aspects of life, and if a philosophy is concerned with mathematics education, the examples must be relevant to that education. The examples become aids to clarification: the philosophy is in the examples, and the examples are in the philosophy. The intention is to provide meaning to an interpretation and to do this in such a way that not only the meaning of critical mathematics education becomes obvious, but also its importance.

I have already used the expression 'critical mathematics education' three times in this introduction without trying to clarify the meaning of 'critical'. However, I shall not try to explain this term in the introduction – that job belongs to the chapters that follow. The whole task of this book is to develop a meaning of 'critical mathematics education' and, by so doing, to prepare the ground for 'a philosophy of critical mathematics education'. May I just state that one dimension of my critical approach is to find a perspective from which mathematics education can be interpreted as part of the language of our technological culture and, consequently, as a part of this culture. Further, it is to look for possibilities in educational practice from which to take such a perspective into consideration. Let this be an introductory remark with respect to the idea of 'critical'.

CHAPTER 1

CRITIQUE AND EDUCATION

In 1966 Theodor Adorno published the article 'Education after Auschwitz', which has been conceived by many as a pillar of critical education.[1] Although it contains several odd formulations, which for instance reveal Adorno's bias against rural perspectives, it has provocatively highlighted the importance of education, not only as an effort to deliver information, but also as part of a cultural and political struggle. Adorno puts education in the position of a social and political force by maintaining that the very first demand of education is that an Auschwitz will never happen again. Even if it were an illusion to believe that education could prevent social and political catastrophes, education cannot set aside the responsibility of trying to fight for human rights. If not, education runs the risk of being a follower.

'Education after Auschwitz' begs an interpretation of 'critique' as an educational concept. As long as 'critique' and 'education' are separated, education easily takes the form of 'delivery of information' or 'socialising young people into an existing culture', but such interpretations become problematic if critique and education are integrated. To suggest such a possibility is to take a huge step, because those two concepts have different origins and have been separated in Western cultural and philosophical tradition.

When used in connection with 'education', the meaning of 'critical' needs some explanation. Some origins of the meaning are found in the *Critique of Pure Reason* in which Kant tries to explicate the general and transcendental conditions for obtaining (certain) knowledge. Via the work of G. W. F. Hegel, critique gets a materialistic interpretation in the hands of Karl Marx. A critique must try to clarify the economical and political conditions for the development of ideas. The term undergoes further evolution in Critical Theory, as developed by Max Horkheimer, Theodor Adorno, Herbert Marcuse and others belonging to the Frankfurt School. Research at the Institut für Sozialforschung in Frankfurt should, as expounded by Horkheimer in his inaugural address in 1931, takes place as interdisciplinary study including philosophy, sociology, economy, history and psychology to interpret social life with the aim of finding opportunities for radical social improvements behind the bulwark of facts.[2]

A further development of Critical Theory is undertaken by Jürgen Habermas, who in *Knowledge and Human Interests* unfolds the theory of the existence of different knowledge constituting interests. Most importantly, he generates the idea that the social sciences must be founded on an interest in emancipation, if they are not to fall into the trap of logical positivism, thus restricting research merely to the activity of outlining correlations between well-defined phenomena. By so doing, social sciences will be colonised by the technical-manipulative research paradigm, according to Habermas. It is not possible to find any platform of neutrality. Social sciences must be 'committed'. A pretended neutral registration of facts will result in an acceptance of the social status quo. Later, Habermas takes a big step beyond the original ideas of Critical Theory in outlining a theory of communicative action.[3]

To gain some idea of critical education, I shall comment on the constructs 'crisis', 'critique' and 'emancipation' and then introduce some of the questions raised by the very use of 'critical education'. The aim is to clarify and to provide concepts with meaning, not to prove that this or that is the meaning. The concepts used in this book all have an open texture. This does not mean that they have no texture, but the reader should not expect them to fit together like pieces of a jigsaw puzzle.

1. CRISIS

We have passed 1984, but we are not situated in a society with the thought police described by Orwell. However, even if we have escaped that terror, we find ourselves in an industrialised and mechanised society stamped by crises and conflicts. We face conflicts because of an unequal distribution of goods, both locally and globally. We face conflicts because of differences in social and economical opportunities. We find social suppression caused by certain ways of handling power structures in society. We witness tensions between the power centres and people without power; between black and white, and between rich and poor. Conflicts exist between western countries and the so-called 'Third World' countries. We face class struggles, even after the belief in a Marxist definition of classes has lost supporters. It is impossible to describe these conflicts in a straightforward way. Conflicts are built up from very different ingredients; only in theory and on (this) paper do they exist as bi-polar antagonisms.[4]

A basic assumption in what follows is that society is fraught with conflicts and crises. I shall not make any attempt to specify the type of society I have in mind. I only maintain that the discussion is exemplified by that western technological society in which I live and about which I know. I shall refuse to interpret society as a well-balanced structure. Society and relations between societies form non-homogeneous patterns. The picture of society as approaching an equilibrium cannot be realistic; such pictures have been painted by liberal economists like Adam Smith indicating that if free enterprise prevailed then harmony would be achieved. My thesis is that whether we look at economy or other social features, no equilibrium is in sight. It is not possible to imagine a balanced society as outlined by Plato in *The Republic*, maximising the good way of governing, because the 'wisest men', i.e. the philosophers, are put in power, and, consequently, no further improvements can take place. The society will become stable, but only in *The Republic*. The critical nature of society becomes a defining aspect of it, and not just a circumstance which we can imagine may disappear by means of some suitable development.

How do we know that conflicts exist? This is not an assumption I shall try to prove through a lengthy argument but I do not find it a haphazard assumption to begin with. If anybody wants, however, to find a proof, it could follow the line of proof of the existence of the external world presented by George Moore.[5] A classical philosophical question has been: If everything we are able to perceive and to know about is what we register by our senses, how shall we ever come to know about the existence of a reality behind these sense-data? How could we know the existence of the external world to be the cause of what we perceive? However, Moore claimed to have solved the problem. He gave a proof which he found could settle the philosophical confusion. He raised his right hand and then his left hand and asked if everybody present could see them; and because the answer was 'yes' he claimed to have completed his proof. What other meaning of any talk about the external world could there be? Moore wanted to re-establish commonsense in the midst of theoretical philosophy and to prevent its degeneration into gentle futility. To deny the validity of the argument and to continue philosophically pondering the question would, according to Moore, be an overwrought diversion away from commonsense, a diversion into an affected language, quite different from the language in which the philosophical problem itself develops. In addition, Moore's proof has the important characteristic, fundamental to every mathematical and formal proof, that it can be repeated as often as needed by anyone wanting to check the proof.

Do conflicts and crises exist in our society? This question could be answered in the same simple and straightforward way. The question does not lead to futile arguments. The proof is everywhere: take a look around and you will see the meaning of conflict and misery. It is contrary to commonsense and to the normal use of language to deny that horrifying situations exist. I conceive the evidence of the critical nature of society as of the same logical nature as Moore's proof. If we have some new philosophical whim and come to doubt whether the external world in fact exists, we need only to hold up our right hand and ask a person next to us if he or she is able to see it. Similarly, if we come to doubt whether the wretchedness in fact exists, we need only to repeat the proof by taking an extra look. Also this proof can be repeated as often as needed. The external world exists, and it is in a miserable state.

I can use terms like suppression, conflict, contradiction, misery, inequality, ecological devastation and exploitation; and it is impossible to deny the relevance of such terms. I shall use the term crisis to include all these phenomena, although I shall continue to use the other concepts as well.[6] They belong to a semantical framework from which it is impossible to escape.[7] 'Crisis' is interpreted as referring to actual, as well as potential, crises. As we shall see, 'crisis' is most useful for exposing the semantical roots of 'critique', but I find 'conflict' more useful when trying to express some of the ideas of critical education in a more straightforward way.

2. CRITIQUE

How does one react to a critical situation? I shall try to consider the question philosophically, and I shall return to the semantic origins of 'crisis' and 'critical' in Greek; here we can trace a connection between the terms. Both notions relate to the Greek words *krisis* and *kritikos*, and in those terms different meanings are included. We find a reference to an activity of making a decision and a judgement – we have to find some criteria to guide us in getting out of a dilemma. We also find a reference to a medical or clinical interpretation: when an illness has reached a turning point, the patient is in a critical situation. Finally we find in ancient times a reference to the idea of interpreting a text, a *kritikos* becomes a philologist.

These interpretations were handed down through antiquity. The text, interpreted by the critic, becomes the Bible or another holy text. The work of a critic became important by its relating the text to its original sources. Texts, and especially religious texts, had, so to say, degenerated through the long-running activity of being copied and recopied, but by careful study of sources this textual distortion could be remedied. This became a critical activity. This activity however could take a new turn by revealing some (perhaps) widespread misunderstandings. A critic might have to confront authorities with basic mistakes, perhaps mistakes essential to the 'myth' of the authority. The term criticism acquired a new connotation.

Critique as a reaction to misunderstandings was emphasised during the period of Enlightenment. A strong belief in the ability of the human spirit was emphasised. Enlightenment became a powerful enterprise, and we reach the age of *Dictionnaire Historique et Critique* (1695–1697) and the *L'Encyclopédie* (1751–1765): We are in the age of Pierre Bayle, Jean Le Rond D'Alembert, Francois de Voltaire and Denis Diderot. They elaborated criticism with the idea that misunderstandings, superstitions and myths can be eliminated if we stop believing authorities and start to trust reason. The activity of getting rid of misunderstandings became a 'critical' activity. Enlightenment could be dangerous for the authorities; however the Enlightenment, in its first phase, was not developed as a political movement and only later did criticism acquire a political dimension.

In the spirit of the Enlightenment, critique, in the form of reason, becomes a weapon in the struggle against superstition. Critique and the use of reason become synonymous. Through the work of Kant this weapon also becomes the object of a critique, although it is still the subject who has to carry out this critique. This idea is nicely expressed by the ambiguity of the title: *Critique of Pure Reason*. The analytical task took the form of a self-critique. Kant had been provoked by David Hume's scepticism, and he tried to outline the conditions for obtaining knowledge. He could not base his investigation on empirical facts, because the concern was to specify the conditions for capturing such facts. Therefore, he had to set out the general and transcendental conditions for obtaining (certain) knowledge in an analysis prior to any empirically based investigation. Thus, critique becomes an epistemic concept, having to do with pointing out conditions for knowledge. This aspect of critique becomes essential for the theory of knowledge.

As mentioned, via the work of Hegel critique undergoes a materialistic interpretation in the hands of Marx. Now a critique does not mean only a critique of a theoretical structure, of a theory of economics for instance, it also comes to mean a critique of the very object of the theory. Critique obtains a double orientation – towards some opinions as well as towards some real situations. Marx delineates a precise interpretation of those aspects of life he found important to criticise. I shall not try to repeat his interpretation, because it is not directly relevant to my interpretation of critical education, but we have to keep in mind the idea of critique as oriented towards both opinion and reality. That leads us to return to the construct of 'crisis'. A 'crisis' is a metaphor for a situation to which to react through the medium of critique. However, if we are not to let critique degenerate into pure activism, we have to protect the double orientation implicit in the notion of critique. Critique still refers to the activity of judging and of finding a way out of a dilemma, and I want it to keep the connotations it derives from analysis, evaluation, judgement and assessment as well as those it derives from 'action'.

A crisis may develop into a critical situation. We have to react. Here we find the roots of a *critical activity* as I want to interpret the term. To be critical means to draw attention to a critical situation, to identify it, to try to grasp it, to understand it and to react to it. These are aspects I shall try to preserve when the concept becomes part of the notion of critical education.

Different possibilities exist for disguising a crisis. Some belief systems serve the function of explaining that an inequality in fact is a necessity. They explain it, so to say, as part of the natural order. A belief system which tries to eliminate a crisis from the common consciousness by establishing a conceptual blindness is sometimes called an ideology.[8] To indicate what is meant by ideology we could think of how differences in the treatment of people because of differences of colour of skin have been explained as appropriate. Some of these ideological structures are condensed into expressions like 'the white man's burden'. They have been fundamental in transforming the phenomenon of colonisation into a humane act. To colonise has been interpreted as a praiseworthy activity of civilisation and to colonise may then be seen as being next to benevolence.

Ideologies hide or disguise conflicts and therefore they tend to reinforce established power structures of society. It has never been urgent for power centres to undermine ideologies which explain how the established social order in fact copies the intrinsic order of things. Ideologies can serve the same function as the 'thought police' in *Nineteen Eighty-Four*. According to Orwell, it is necessary to control the thought of people if suppression is to be sustained and a society based on inequality be preserved. It may be right that it is necessary to control the imagination of people, but wrong to assume that this has to be performed by 'thought police', i.e. by a brute act of power. The control can be much more sophisticated and 'humane'. It can be carried out by ideological structures generating a grammar for what it is possible to think and what is not. Then people 'preserve their freedom' to think whatever they want but within limitations which are difficult to identify and therefore to recognise as being restrictive.[9] A kind of Newspeak may serve as a very efficient 'thought police'.

A critique of ideology is directed towards certain belief systems and attempting to do this in a theoretically based and more organised way is what characterises a *critical theory*. The Frankfurt School indicates one possibility for realising this. According to the positivistic research paradigm, the only thing possible to deal with in research are facts, and the objective of research is to identify correlations between different sets of facts. Research has only to do with what is actual. But that is not the only aspect involved in critical theory. To be critical means to be directed towards a critical situation and to look for alternatives, perhaps revealed by the situation itself. It means to try to identify possible alternatives. Positivistic research looks for what is actual; critical theory looks for what is possible in light of what is actual and critical. It is not possible to identify crises by any objective method. This means we have to give up any attempt to find a universal technique of critique (but not to give up being critical). A critical theory needs interdisciplinarity and it may well be important to look outside science and scientific approaches. As mentioned earlier, Horkheimer has emphasised the condition of interdisciplinarity in all critical studies and especially in a critique of ideology.

We assume that critical situations exist even if we do not possess general methods for their identification. The assumption is that crises cannot be reduced to mental phenomena. This means that critical activity is not just a deliberate undertaking, it has its reasons outside the fluctuating fashions of mind. Crises belong to reality. They are not merely conceptual constructions, meaning that a different way of looking at reality may create a picture we can appreciate. We can make society look homogeneous by introducing a suitable belief system, but reality as such is still critical.[10] Therefore, I do not accept the relativistic approach which suggests that the existence of conflicts is only a question of interpretation. But how do we know that the identification of crises is not just a question of choosing certain value systems? Again, my best suggestion is to go back to the commonsense proof for the existence of crises.

When crises have an objective status they can interact. One conflict can influence another. Causal relationships between crises may exist. According to (a simplified version of) Marxism we may assume that a hierarchical structure of crises exists with one crisis being fundamental, namely the capitalistic exploitation of the working class. From that one crisis all other sorts of social misery appeared to be explained; and after solving the class struggle we are asked to follow Marx into a harmonious society. Utopia becomes real. Therefore, the whole enterprise of social development must concentrate on solving this fundamental crisis. To be critical and radical then means to support the working class in its class struggle. To try to do something different could even be condemned as a reactionary divergence from the 'true course'.

I find this interpretation of critique to rest upon a strong but incorrect assumption about the existence of a hierarchical order of crises. I find the situation much more complicated: crises may interact in a chaotic way. We may talk about the 'wild life of crises'. Crises may interact, but it is impossible to outline general features in the way this interaction takes place. It is not possible to presume that a solution to class exploitation will imply a solution to other critical features of society. We cannot presuppose that a classless society is without crises.[11] New catastrophes can always emerge on the horizon. They need not be consequences of some fundamental problems we already know about. The emerging crises may interact with problems by which we are already surrounded. Therefore, I see the interaction of crises as belonging to the dynamic structure of societies. Social change may be a reaction to actual or potential problems.

When we negate the postulate about the structural order between crises, we also negate the idea of a determinism existing between the social phenomena implicit in Marx's discussion of social development. A fundamental theoretical assumption behind Marx's economical analyses is that it is possible theoretically to identify a trend in social development, due to the fact that it is possible to identify mechanisms in economical development. At this point Marx was inspired by the scientific paradigm of Isaac Newton. Marx hoped to be able to formulate some fundamental laws of the transformation of capital of the same level of generality as natural laws, which could explain the transformation from one social structure to another in terms of a logic of capital. These laws of development are assumed by Marx to be of a deterministic nature. They contain no stochastic elements but are bound to the realm of necessity. Marx develops his theory within the borders of determinism.

These borders, however, are problematic for the development of the concept of critique, because they create heavy restrictions for what it means to be critical by assuming a deterministic order of critical situations. They also produce the well known technological optimism of Marxism because the development of technology sets up the basic conditions for economic order and social development: the future is already contained in the present, and a socio-economic revolution must take place. I do not assume any mechanical relationship between crises and I do not rely on any optimism in technological development, nor on any pessimism: the situation is open-ended. Crises may interact because they belong to reality, but because the interaction is stochastic it is impossible to predict the consequences of these interactions, and therefore it is impossible to predict the consequence of any 'management of crises'. When we leave the deterministic paradigm, it becomes impossible to outline main trends in social development. Marxist expressions like 'the necessity of history' will lose their meanings.

When we negate the possibility of a deterministic relationship between crises and the general hierarchical structure of crises, we also negate the possibility of describing what a genuine critique may consist of, and it becomes impossible to rely on any materialistic one-way-ethics, which determine the value of an action according to whether it supports the class struggle or not. When the hierarchical structure breaks down, such ethics become dogmatic. Because we are unable to make predictions, the situation becomes stochastic and taking any action means running a risk. That is the condition for ethical commitment and for every form of critique. Critique becomes an open-ended activity.

3. EMANCIPATION?

The goal of a critical activity has sometimes been described as an *emancipation* but this (woolly) construct has different meanings. An emancipation could be the result of a critique of ideology, and that means a freedom from stereotypes of thoughts. The freedom is similar to the escape from a neurotic state of mind obtained when a person leaves the psychoanalyst's couch after successful treatment, and therefore emancipation creates a condition for resistance towards a threatening world.[12] The other aspect of emancipation is the actual liberation from material constraints, like the freedom the slaves got when they left their owners as free human beings when slavery was abolished.

The widespread use of emancipation highlights the 'naive' aspects of critique. The notion of emancipation indicates the existence of a possible result of a critique. I find this problematic and to a great extent I shall refrain from using the term 'emancipation', although it has taken up a central position in the vocabulary of critical education. When we react to one crisis our action influences other crises as well. Some may disappear. Some new ones may emerge. New schemas for interaction between crises will emerge. Therefore, emancipation cannot have any absolute references, but the use of the word 'emancipation' may easily imply that some ideas can be interpreted as being definitely wrong – and so may be discarded by the emancipated person. However, such absolutism is not a becoming idea to my epistemology.

A 'solution' to a critical situation could consist of a transformation to a new but future, and therefore (only) potential, crisis – often however with a more dramatic risk structure. For example, we can solve the conflicts emerging because of lack of energy by introducing nuclear power. A crisis which then faces us in the distant future is a nuclear breakdown. A second Chernobyl is always a possibility. Some economic problems in industry could be solved by speeding up production, but this could also mean a reaching the limits of nature more quickly. Ecological problems often seem to be created by taking more risks in production. This has some short term (economic) advantages, and the implicit argument in favour of taking the risk is that the catastrophic situation need not happen but is only a possibility. And this is true. This underlines how difficult it is to analyse the consequences of ways of handling specific crises. We do not face calculable consequences but structures of risk, and these structures may interact in the future in unpredictable ways. This does not mean that it is impossible to analyse anything or that we cannot talk about good and bad ways of reacting to a conflict. The point is that it is impossible to outline a full picture of the situation. Our discussion will always be based on simplified and partial notions of some scenarios – and we cannot know if everything of main relevance for the analysis has been taken into consideration. We are condemned to live in a 'risk society'.

As two extremes, we could talk about black and white dialectics of crises, reminding ourselves that the normal colour is grey.[13] In the black dialectics all ways of handling a crisis result in a more chaotic and terrible situation. Any attempt to solve the problem results in its transformation into a more complex and dangerous situation than it originally was. In the black dialectics we are in a labyrinth with no way out, every step leading deeper into the labyrinth. The white dialectics describe the opposite situation. We have to remember that we do not have any means of determining whether we are situated in light grey or dark grey dialectics. In fact, the metaphor of the labyrinth has to be changed a little. We are situated in a labyrinth, with lanes continually changing and rearranging depending on each step we take. The structure of the labyrinth changes depending on our attempts to get out of our problematic situation. Therefore, the notion of emancipation becomes misleading.

From the existence of a not-totally-black form of dialectics we may develop the 'principle of hope'.[14] If crises are organised in a hierarchical structure, with capitalistic exploitation as the main cause for social misery, it does not make much sense to try to solve 'low level problems'.

The rational act is to wait until the class struggle has been brought to victory, and until then all critical activity must pursue that aim. Revisionism becomes ridiculous because it leaves fundamental problems unsolved. But because the main problem is so fundamental and the solution so difficult to imagine, the consequence easily becomes fatalistic. Class struggles are not going to be solved in our time – that was the perspective of most Marxists. This fatalism is not the consequence of the grey dialectics assumption. In this case any sort of action addressing critical situations may be useful. Our labyrinthic situation may be improved even if we cannot know for sure. Utopia is lost, but now even minor changes make sense. With grey coloured dialectics it becomes difficult to establish guidelines for critical practice – and this freedom (confusion) has arisen from negating the thesis of hierarchical order. It becomes difficult for me to see any reference-points for 'emancipation' outside the perspective of a day-dreamer, but the principle of hope can in theory be preserved, even in education.

4. CRITICAL EDUCATION

Which institutions in society ought to react to the critical nature of society? Which institutions are able to do so? Adorno suggests educational institutions have this role, and this means we have to find a meeting place for 'critique' and 'education'. Since the 1960's it has been a main concern in critical education to constitute education in line with the programme of Critical Theory, but no such unifying programme has ever been outlined. The label 'critical education' has been used in a loose and non-committal way to refer to a variety of educational ideas like: Try to establish the conditions for emancipation! and: Build upon the interest in emancipation! However, such maxims could easily be interpreted in different ways. What does it mean to relate an educational practice to the interest in emancipation, and what does it mean to develop educational research which follows the analyses of Habermas concerning the knowledge constituting interests?[15]

Critical education has manifested itself in a variety of catchwords: problem orientation, project organisation, *Fachkritik*,[16] subjective relevance, emancipation, etc. Elements of a critical education are found in the works of Oskar Negt, who discusses the principle of exemplarity, according to which it is possible to obtain a general understanding of a subject focusing one's research on a specific topic, Klaus Mollenhauer, who discusses education in relation to questions concerning democracy, Wolfgang Lempert, who conceives of education as being similar to a critique of ideology, and Knud Illeris, who elaborates an educational theory aiming at problem orientation and project organisation.[17] The concept of education connected to an emancipatory interest has different sources from Critical Theory.[18] Paulo Freire, for instance, develops his ideas quite independently of that framework.[19] Another source of inspiration is found in 'Geisteswissenschaftliche Pädagogik' which is inspired by hermeneutics as worked out by Wilhelm Dilthey. Martin Wagenschein is yet another educationist who must be mentioned.[20] He discusses the principle of exemplarity which has been one of the main ideas in the project organisation of education.[21]

If we tentatively try to summarise critical education by stating one simple idea it could be: *If educational practice and research are to be critical, they must address conflicts and crises in society.* Critical education must disclose inequalities and suppression of whatever kind. A critical education must not simply contribute to the prolonging of existing social relationships. It cannot be the means for continuing existing inequalities in society. To be critical, education must react to the critical nature of society. (Logically speaking, if somebody could imagine a society with no actual or potential conflict, a society with everything in the 'right order', critical education would be superfluous. If Plato's absolutely perfect society existed and was indeed so, it would not be necessary to invent a critical education. We would not expect a section about critical education in *The Republic*. In that case education would comprise a simple introduction to the life and institutions of the ideal society.) Critical education is brought into existence as a result of the critical nature of society – and some of the perplexity of critical education is implied by the (necessarily) fragmentary analyses of crises and conflicts.

Why, then, should education react to critical situations in society? There is no simple answer to this question. The question is not of the same nature as: Do social crises exist? We cannot find any trivial proof by the raising of Moore's right hand. We could try to find several reasons, but I shall not attempt to find a persuasive argument. The question: Why should education react to the critical structure of society? is not a rhetorical question. For instance, it could well be argued that it is important to 'protect' students from the misery of society. What I shall try to do is to concentrate on providing a more precise meaning to the expression 'critical education', rather than on arguing that it should be a global activity. I hope to show that 'critical education' is a meaningful concept in today's society.

What does it mean for education to be related to critical situations in society? If we, together with Marx, assume the existence of a hierarchical order of crises, the simple answer is that education must try to constitute a force in the class struggle. Traditional Marxist educational theory has followed this paradigm: in a socialist society the need for critical education does not exist. As in the Platonic state, education then comes to mean the introduction of students into the harmonious state, an idea which became a disaster for education in the Eastern Bloc. A different situation emerges when a non-dogmatic interpretation of crises and conflicts is adopted. By my interpretation critical education must be a reaction to all sorts of critical features of society. It has to take into consideration not only class inequalities but other inequalities as well. Critical education could take the form of anti-racist education or be directed towards the difference in treatment of boys and girls at school, it could try to attack elitism, etc. The existence of a wilderness of crises and grey zones of dialectics are the presumptions of this approach. Education is free to interpret its contributions to the reconstruction of our lifeworld.[22]

Education can react to the critical nature of society in different ways. The school system is embedded in a society fraught by crises, some of which themselves are manifest in the school. Inequality in society creates differences in opportunity at school. Society is the background for the structuring of schooling, which therefore cannot merely be governed by well-based educational principles; instead the school itself is part of the political battlefield. Highly technological societies with a social and economical structure such as that in my country, need a workforce with diverse qualifications. An equal distribution of qualifications is not viable; on the contrary, the society needs a polarisation of qualifications. A proportion of the work force must possess high qualifications but a certain proportion also has to accept the drudgery jobs, which have the lowest salary. It seems a condition for the preservation of the economic system in capitalist countries that some part of the labour force is kept submerged but ready to fill the gaps that emerge as a result of the new needs of society. Unemployment is 'useful'. Our society 'needs' an unequal distribution of qualifications, and this puts pressure on schools to produce this state of affairs. Inequality outside school is therefore going to be reflected inside school. Elitism, conceived as the idea that only a certain proportion of students should get a higher education based on their better disposition to learn, conforms with this structural demand of society. But elitism turns reason upside down by relating the conditions for further education to the intellectual potential of the individual student. This could be summarised by the thesis that schooling leads to the reproduction of social structures. This reproduction includes a reproduction of the division of labour, a reproduction of the distribution of power between the individual and the state and between social groups, and finally a reproduction of traditional cultural values. In short, the critical aspects of society are part of life in school. A critical education must seek to respond to this.

Schools produce knowledge, routines and competencies, as well as support ideological beliefs. If an education is to be critical, it has to take into consideration the critical background of schooling and to try to develop possibilities for an awareness of conflicts as well as to provide competencies which are important for dealing with such critical situations.[23] In addition, we have to discuss to what extent it is possible to limit educational practice to such aims. It may well be the case that to be a properly functioning critical education, the total educational enterprise has also to relate to features that are different from the critical structures of society. A main question concerns the relationship of traditionally established areas of knowledge: What is the status of the traditional school subjects in a critical education?

5. LITERACY AND MATHEMACY

In *Schooling for Democracy* Henry Giroux gives the following formulation of the idea of critical education: "Schools need to be defended as an important public service that educates students to be critical citizens who can think, challenge, take risks, and believe that their actions will make a difference in the larger society. This means that public schools should become places that provide the opportunity for literate occasions, that is, that provide opportunities for students to share their experiences, to work in social relations that emphasise care and concern for others, and to be introduced to forms of knowledge that provide them with the conviction and opportunity to fight for a quality of life in which all human beings benefit."[24] Here it is emphasised that school must educate students to become *critical citizens*, prepared to take risks and to challenge and to believe that their actions can make a difference in the larger society.

In critical education the discussion of 'literacy' has played a major role, especially as a result of the work of Freire who developed the political dimension of education from that term. The work of Freire is important because it starts with the basic assumption that education has to relate to critical structures of society, and because it shows how to interpret this assumption with respect to an educational practice which also teaches people how to read and write.

Literacy has been described as a double-edged sword. Literacy is necessary in today's society for informing people about their obligations, and for people to be employable in essential work processes. However, literacy can serve the purpose of empowerment because it can be a means of organising and reorganising interpretations of social institutions, traditions and proposals for political reforms.[25] Literacy is not just a competency for reading and writing, an ability which can simply be tested and controlled, but it also has a critical dimension. Because of its potential for reorganising human interpretations of reality, literacy can be seen as part of a critique of ideology (following Lempert's suggestion) and it can become a means of pinpointing inequality and suppression and therefore a tool for identifying the critical features of a society. Giroux formulates it in this way: "... literacy as a radical construct has to be rooted in a spirit of critique and project of possibility that enables people to participate in the understanding and transformation of their society. As both the mastery of specific skills and particular forms of knowledge, literacy had to become a precondition for social and cultural emancipation."[26] This formulation connects literacy with the 'spirit of critique'. The goal is not only a better understanding but also the transformation of society.

It is part of the purpose of critical education to secure an equal distribution of opportunities like reading and writing. But that is not the only thing related to literacy in a critical education. Giroux underlines the fact that "literacy is not just related to the poor or to the inability of subordinate groups to read and write adequately; it is also fundamentally related to forms of political and ideological ignorance that function as a refusal to know the limits and political consequences of one's view of the world. ... What is important to recognise here is the need to reconstitute a radical view of literacy that revolves around the importance of naming and transforming those ideological and social conditions that undermine the possibility for forms of community and public life organised around the imperatives of a radical democracy."[27] A new term, 'democracy', is introduced and an important discussion can be expounded about education based on this concept. In addition, it is emphasised that critical education has to do with more than just teaching illiterate people basic skills. Naturally, this is an important job of education, but it cannot be identified as the (main) characteristic of critical education. Literacy has also to do with the general interpretation of social life. Critical education must fight against ideological constraints. It must react to the conflicts and differences in opportunities which society thrusts upon schools, and it must provide competencies which enable people to confront the critical nature of society.

Giroux also finds that literacy relates to emancipation: "An emancipatory theory of literacy points to the need to develop an alternative discourse and critical reading of how ideology, culture and power work within capitalist societies to limit, disorganise, and marginalise the more critical and radical everyday experiences and commonsense perceptions of individuals."[28] This is a variation on the formulation of the thesis that an ideology hides crises and conflicts. Ideologies are built into our basic grammar and therefore literacy can perpetrate an attack on such belief systems. Literacy is part of the project of developing an alternative discourse, which is important because ideology, culture and power may be able to 'limit, disorganise and marginalise' critical and radical everyday experiences.

At this point Giroux is able to reach a more general conclusion: "Literacy in this wider view can serve to empower people through a combination of pedagogical skills and critical analysis and also function as a vehicle for examining how cultural definitions of gender, race, class, and subjectivity are constituted as both historical and social constructs."[29] Literacy now means empowerment. It may develop conditions for human beings to locate themselves in history and recognise their position in society and, by doing so, enable human beings to function in society. People become not just onlookers but also actors. Through literacy we come to know about our position in the world, which is exemplified most directly in the practice of Freire. Oppressed people become able to participate in the struggle for improving their opportunities.[30]

Could 'mathemacy' be substituted for 'literacy' in the considerations in the previous section?[31] By mathemacy we may initially mean 'ability to calculate and use mathematical and formal techniques', although we may hope later to provide a more elaborate differentiation of the concept. Could mathemacy be conceived of as a double-edged sword? Could mathemacy also be used for the purpose of empowerment? Could mathemacy help people to reorganise their views about social institutions, traditions and possibilities in political actions? If it is possible to substitute 'mathemacy' for 'literacy', then it makes sense to talk about a critical mathematical education at least to the same extent as it does to talk about critical education with literacy as a focal point. However, it is not obvious that the competencies of literacy and mathemacy have parallel roles to play in society. If language constitutes a part of our world view, and therefore a part of our world, it seems to follow that a critical education should accentuate literacy. But what does mathematics constitute?[32]

Reformulating another of Giroux's propositions on literacy we may say: Mathemacy, as a radical construct, has to be rooted in the spirit of critique and the project of possibility that enables people to participate in the understanding and transformation of their society and, therefore, mathemacy becomes a precondition for social and cultural emancipation. Could this be more than a half-empty assertion? What might the meaning of mathemacy be, if it is to fit into this formulation? Freire expands the notion of literacy such that it includes more than just reading and writing abilities: what sort of extension of mathemacy is needed? It was emphasised that literacy is related not just to the inability of subordinated groups to read and write adequately. Is the same the case with regard to mathemacy? Is mathemacy also related to the forms of political and ideological ignorance that function as a refusal to know the limits and political consequences of one's view of the world? Could mathemacy be involved in actively naming and transforming those ideological and social conditions that undermine the possibility for forms of community and public life organised around the imperatives of a radical democracy?

It would naturally be too simple to take as being axiomatic the proposition that mathemacy has a role to play in society similar to that of literacy. This may be the case, but differences and similarities have to be analysed and discussed before we can try to provide the term 'critical mathematical education' with more meaning. The problem is, in fact, that some of the basic terms which have been used in outlining critical activity and in specifying the general assumptions in critical education, need not be operative in an attempt to identify a possible meaning for critical mathematical education. A basic term is 'crisis', but what does that mean in relation to mathematics? What does 'critique of ideology' mean? These concepts have been operative in establishing literacy as part of critical education, but they need not be operative in establishing a critical mathematics education. My intention is, therefore, not to extrapolate from the conceptual framework set up in this chapter but to use it as it has already been used: to provide inspiration for the further development of critical education.

In his concluding remarks Giroux talks about radical democracy and relates this to the idea of critical education. Many educators have reflected on democracy. This has been the case since the emergence of the idea that education should be public mass education and not just aimed at the elite. I believe that a careful look at ideas related to democracy could reveal something more about the nature of critical mathematics education. A further step will be to look more carefully at the nature of the role of mathematics in society. Does mathematics have anything to do with critical structures of society?

CHAPTER 2

DEMOCRACY AND EDUCATION

When education becomes not only an enterprise for the elite but 'mass education', it also becomes related to democratic ideals. This was pointed out at the beginning of this century by John Dewey: "Since a democratic society repudiates the principle of external authority, it must find a substitute in voluntary disposition and interest; these can be crated only by education. But there is a deeper explanation. A democracy is more than a form of government; it is primarily a mode of associated living, of conjoint communicated experience."[1] Dewey more than any other has unified the discussion of education and democracy. His contribution is vital to the ramification of educational discussion – as vital as the contribution of Adorno, i.e. the integration of a critical dimension.

'Democracy' is a concept broader than that of 'critique', and yet 'democracy' is sharp enough to illuminate aspects of education not made visible by 'critique'. Therefore, I find the contributions of both Dewey and Adorno important to education; and in the final section of this chapter, I shall try to describe a term which can bring together aspects of both perspectives in such a way as to reach a better understanding of 'critical education'. As already noted, the two perspectives are related: Giroux used 'democracy' in his discussion of critical education.

It is generally agreed that democracy is a most attractive feature of society, but at the same time widespread disagreement exists about what democracy means.[2] Democracy refers to a bouquet of different ideas, hopes and utopias. Therefore, although it is impossible to pinpoint any simple definition of democracy, we can try to grasp the concept by outlining some ideas which relate to it.[3]

1. LINKS BETWEEN 'DEMOCRACY' AND 'EDUCATION'

Democracy refers to at least the four following considerations: The first of these concerns the formal procedures for electing a government. However, formal procedures have to be specified not only for election but also for ruling, for the distribution of power, for judging – in short, for the interplay between the different institutions of a democracy. Second, democracy presupposes a fair distribution of social services and goods in society, such as welfare, education, hospitals, etc. Consequently, a substantial part of the theoretical analysis of democratic ideas is concerned with the types of goods and facilities which have to be distributed in a fair way. We cannot call something a democracy, based on votes and elections, if it always turns out that the majority decides that the minority should not benefit from any of the social services. A democracy has to be fair, but what is the interpretation of 'fair'? Third, democracy assumes equal opportunities and obligations for every member of society. There must be no differences in opportunities based on differences in social background, sex, colour of skin, etc. Everybody must be treated equally under the law, and similarly everybody must obey the law. But what does 'equal opportunity' mean? According to the liberal and idealistic tradition, it means the unrestricted possibility for persons to try to do whatever they (legally) want to do; while the materialistic tradition has underlined that it is not enough to decrease the number of restrictions, society must actually provide the conditions for everybody to be able to pursue his or her interests. In this way, discussions about democracy becomes discussion about freedom. Fourth, democracy must permit the possibility for citizens to participate in discussion and evaluation of the conditions for, and consequences of, the governing which takes place. A 'democratic life' is indispensable. Could we imagine a democracy in which nobody actually knew what the (democratic) decisions were about?

To sum up, democracy refers to *formal* conditions concerning the interplay between the institutions of a democracy, *material* conditions concerning distribution of goods and services, *ethical* conditions concerning equality, and finally conditions concerning the *possibility for participation* and re-action.

Although very general, this outline of democracy also points to different links between democracy and education. First, education can include an introduction to democratic life in society. This could, for instance, be interpreted as meaning that school has to teach students to appreciate basic democratic values, such as equality, fraternity and tolerance – education could be a preparation for a democratic life.[4] Second, a focus on democracy can imply a concern for the distribution of knowledge and educational possibilities in society. If the educational system is to represent a democracy, it must provide equal opportunities for the members of that society, including the students. Third, democracy in education can refer to the 'life' of school and classroom, implying that this may represent democratic values. This idea is related to the first aspect mentioned above (education as preparation for a democratic life) in the assumption that a basic way of learning democratic values is actually to participate in a democratic life. Fourth, democracy and education can have something to do with the content matter questions.

Let me expand some of these remarks a little further. A fair distribution of social services implies that in a democratic society every child and adolescent will have equal access to schooling and learning. This naturally leads to a discussion about equality. What does equality in education mean? Clearly, students appear to receive different sorts of education while living in the same society, even in societies which are supposed to be democratic. How is this possible? Some maintain that only students with the same abilities can be treated equally. This may be convenient for practical purposes: it seems easier to teach a group of students who are at about the same level. But in what sense is this in accord with democratic ideals? Several investigations have indicated that in many countries working class children benefit less from schooling than other children.[5] It is also well-documented that differences in achievement correlate in some countries with gender, at least in some subjects. Schools seem to serve the function of reproducing social structures, including the division of labour, the distribution of power both between the individual and the state and among social groups, and, finally, it seems to reproduce traditional cultural values. What does this mean for our interpretation of education by democratic standards? To be in accordance with the ideals of democracy, schools have to react to the different ways in which society reproduces itself, and it must try to counterbalance some of these reproductive forces in order to provide equal distribution of what schooling may offer, including opportunities for further education and vocational life. Mathematics education plays a major role as gate-keeper to access further education; therefore, it becomes important to discuss any lack of equal opportunities that may be produced by this subject.

Education has to do with the content as well as the distribution of acquired competencies. Although I am not going to pay attention to the distribution of competencies here, this does not mean that this is of minor importance. However, I shall concentrate on one aspect of democracy: the possibility for citizens to participate in democratic life. Education can help to develop a democratic attitude among students, although the development of a democratic attitude cannot be the only thing to be developed in school. Democracy is not just a question of adopting appropriate attitudes but has also to do with competencies with respect to participating in democratic processes. Education must try to provide students with competencies which enable them to identify and react to social repression. This aspect is also stressed in some of Giroux's formulations and expressed by his concern for a critical democracy.

The questions I shall address are: What sort of competencies, important for a democracy, can be supported by education? What is the nature of such competencies in a highly technological society? Can mathematics education be of importance in providing a foundation for students' participation in a democratic life as adults? Let me add immediately: My intention is not to provide specific answers but to examine questions. It is the link between democracy and education we shall pursue – yet without assuming that other links are of minor importance.

2. BASIC DEMOCRATIC IDEAS

'Democracy' has its origin in the Greek word *democratia* meaning 'ruled by the people'. *Demos* means 'people' and *kratos* means 'rule'. Ancient Athenian democracy was described in a speech by Pericles (rewritten by Thucydides) and it expresses some of the basic ideas in a democracy: "Our constitution is called a democracy because power is in the hands not of a minority but of the whole people. When it is a question of settling private disputes, everyone is equal before the law; when it is a question of putting one person before another in positions of public responsibility, what counts is not membership of a particular class, but the actual ability which the man possesses. No one, so long as he has it in him to be of service to the state, is kept in political obscurity because of poverty. ... We give our obedience to those whom we put in positions of authority, and we obey the laws themselves, especially those which are for the protection of the oppressed, and those unwritten laws which it is an acknowledged shame to break."[6]

Democracy characterises a means of social control. It is a way of handling power, and, as Pericles points out, in a democracy power must be in the hands of the whole people. The people must rule. However, already in ancient Greek times the concept needed further interpretation. Who are the people? Not only the rich, as Pericles underlines. Wealth does not matter. Nor does the class to which a person belongs. But what about women? And what about slaves? And what does 'rule' actually mean? Even from the beginning the idea of democracy was discussed from a philosophical point of view. And the idea has had enemies and critics. Among the earliest and most prominent critics was Plato.

In *The Social Contract* published in 1762 Jean-Jacques Rousseau makes a classification of different types of government, and he writes: "... the sovereign may put the government in the hands of the whole people, or of the greater part of the people, so that there are more citizen-magistrates than there are ordinary private citizens. This form of government is known as democracy."[7] This definition closely follows Pericles', but Rousseau takes a step further to express explicitly the idea of *direct democracy*. Everybody, whoever 'the people' might be according to Rousseau, should be able actually to participate in the ruling. If we conceive this to be a genuine definition of democracy, it is obvious that democracy is impossible in most modern societies. The concept of democracy has a limited range of application; and only very small and homogeneous societies can be democratic.

If we give up the idea of direct democracy and try to find a more feasible interpretation, applicable to a wider range of societies, we face the problem of the delegation of sovereignty: How is it possible to combine democracy with the necessity of selecting a smaller group of people actually to do the ruling? This question always arises from *representative democracy*. Pericles emphasises that we have to obey the people put in the position of authority. This is an important part of democracy. The necessity of the delegation of sovereignty is also implied by the fact that ruling presupposes specific qualifications that are not of a common nature, particularly not if we have a complex modern society or organisation in mind. The people in charge must have specific knowledge about the domain they are ruling. Perhaps specialised education is needed. I shall accept the delegation of sovereignty as a necessity in what follows. But how is it possible to control the people in charge? This problem accompanies every attempt to extend democracy beyond the exclusive direct democracy described by Rousseau.[8]

Every definition of democracy includes conditions about the nature and organisation of the society in question. Direct democracy, as described by Rousseau, cannot be realised in all societies. A properly functioning direct democracy must take place in a context which fulfils certain conditions. Rousseau specifies the following: "First, a very small state, where the people may be readily assembled and where each citizen may easily know all the others. Secondly, a great simplicity of manners and morals, to prevent excessive business and thorny discussions. Thirdly, a large measure of equality in social rank and fortune, without which equality in rights and authority will not last long. Finally, little or no luxury; for luxury is either the effect of riches or it makes riches necessary; it corrupts both the rich and the poor; it surrenders the country to indolence and vanity; it deprives the state of all the citizens by making some the slaves of others and all the slaves of opinion."[9]

Rousseau finds conditions of a sociological nature to be important. The community must be small, and a high measure of equality must exist. However, even if we abandon the idea of direct democracy, we may expect some assumptions about the social structure to be fulfilled for democracy to be possible. If material in addition to formal conditions are basic, democracy must also be developed through the economical development of society. But if formal conditions have priority, the important thing is to figure out a proper structure for election and ruling; and every society can approach democratic life when such democratic procedures are established.

In spite of the fact that this formal aspect of democracy is important, and even though it is trampled on in societies which claim to be democratic, I shall concentrate entirely on other aspects of democracy. An important question with respect to representative democracies is how the material and the ethical conditions, and the condition about participation, may develop along with the social and economical development of society. Will it be harder and harder to fulfil the material conditions of democracy as societies develop into a greater organisational complexity? Or will the opposite be the case? Does it make any difference if we are dealing with a highly technological society? Does this social dimension raise new obstacles for democracy? What is in fact the content of the conditions for participation in democratic life in this case?[10] A democratically organised society could approach one having an undemocratic structure, even if the members of the society are appreciating and supporting democratic values. Values and attitudes are not enough to ensure a functioning democracy. Social structures may have a defining importance, and therefore the development of such structures becomes crucial for the possibilities for a democratic life; and trends in this development are determined by technological development.

A specific analysis of the material conditions for democracy can be made from a Marxist position, maintaining that no democracy can be established in a society governed by a capitalist economy. Capitalism prevents the equal distribution of goods, opportunities and power, and therefore any talk about democracy in a capitalistic society becomes (nearly) empty. Capitalistic structures therefore must be defeated before any democracy can be established. The means for overcoming a capitalist economy cannot be expected to be of a democratic nature. Marx's thesis is that capitalism does not allow room for any real deviation. Capitalism could try to appear to be a democracy, but this can only be a superficial pretence. The class in power cannot be expected to give away any of its power. This leads, as we know, to assumptions about a temporary period with a 'dictatorship of the proletariat'. This line of thought exemplifies an analysis of conditions for democracy in which certain economic structures become an absolute barrier for a democracy.

I shall not try analytically to add to the economic conditions for democracy until a level is reached which makes democracy in a capitalistic society impossible. But discussion of the non-formal conditions is necessary, and I find it essential to take technological developments into consideration. I do not believe in the idealistic hope of a nearly condition-free democracy.[11] As mentioned, my concern is concentrated on the non-formal conditions and especially on the aspect of democracy having to do with possibilities for participating in a democratic life. This means that I maintain an interest in the Rousseauan idea that democracy has to do with participation.

3. DEMOCRATIC COMPETENCE

We shall make a distinction between the competence which the people in charge must possess if they are to be able to take well-founded decisions and act in an appropriate way, and the competence which is presupposed if people are to be able to judge the results and consequences of the ruling process. We shall distinguish between competence in ruling and a *democratic competence*. Democratic competence is to be ascribed to the majority; it must be presupposed in a functioning, representative democracy. Democratic competence is the basic knowledge and understanding necessary if the delegation of sovereignty is to be subjected to any sort of control. It is a condition for participation and reaction. Interpretations of this competence varies between two extremes: One sees democratic competence as a natural ability of human beings, the other finds it to be an acquired ability.

In what could be called the classical and idealistic interpretation of democracy the basic idea is: While the ruling-competence of the people in charge is of a special nature, the judging-competence is universal.[12] A classical way of arguing for the common nature of democratic competence is as follows: The judgement of an act carried out by the people in charge is a moral or ethical judgement, and ethics are not matters of fact, information or higher education. No ethical statement is implied by conjunctions of factual statements. Norms do not rest on empirical facts or on results of empirical tests. Nor is an understanding of norms restricted to any small group of people possessing certain qualifications. To argue in favour of a normative statement only rationality (or common sense) is needed, and rationality is common in nature. We need not believe any statement unless we realise that it has to be so. No norm needs to be conceived as true simply because authority tells us so. No ecclesiastical authority of morals exists. The only authority we have to believe is our own capacity for thinking. Dogmatism is refused, and that settles the problem about the content of democratic competence.

In *Considerations on Representative Government* published in 1861, John Stuart Mill emphasises that a radical distinction has to be made between controlling a government and actually doing the ruling.[13] Making this distinction, he introduces a discussion about the nature of democratic competence. However, Mill has in mind a distinction slightly different from the distinction between competence of ruling and democratic competence. He underlines the point that a certain body of persons may be able to control 'everything', but that they cannot possibly do everything. Further, Mill remarks that in many cases the actual control will be more efficient the less the ruling body personally attempts actually to do the ruling. The distinction Mill has in mind is between the government and the Civil Service, and he is concerned about government's control of the actual ruling. This exemplifies a part of the problem of democratic control: How is it possible for the representatives to control the actual ruling?[14] However, a more fundamental question is: How is it possible for the people in general to control the representatives and the actual ruling? How is it possible to participate, not in the actual ruling, but in democratic life, including discussions and evaluations of the ruling? The nature of democratic competence seems to be a problem. But basically Mill is optimistic about the actual rational behaviour of human beings, even if their capacity might be limited. In that sense he trusts rationality.

In considering Mill we have approached the idea that the existence and nature of democratic competence is problematic. We cannot presuppose that democratic competence automatically exists. It has to be developed, but by which institutions in society? Does this imply that we assume the existence of some sort of evaluative authority? Does it make sense to accept the delegation of sovereignty but not to accept that people are sovereign in their evaluation of the result of this delegation? A disbelief in the soundness of democratic competence could easily turn into dogmatism: We have to teach you before you are able to decide properly!

As an extended parenthesis in concluding this section, I shall mention a particular way of 'solving' the problem of the existence and nature of democratic competence. The classical interpretation of democracy draws attention to the question: How do we deal with ruling? The actual election of the people in charge has not been the most important issue. Naturally for this election to be democratic, formal rules for election have to be specified and obeyed, but the idea of election has been seen as subordinate to the real concern of democracy: the appropriate way of ruling. It is however possible to turn this upside down and define democratic life as primarily concerned with election and not with ruling. What must be democratic, then, is the election of the government. This makes formal conditions important in a discussion of democracy. The other conditions for democracy become insignificant, and no attention need be paid to the establishment of a democratic competence. This is a possible 'solution' to our problem.

Such a formal interpretation of democracy has been suggested by Joseph A. Schumpeter in *Capitalism, Socialism and Democracy*, first published in 1943. Schumpeter discusses previous interpretations of democracy, and he criticises what he finds to be the basic assumption behind these interpretations: the rational behaviour of people in general. According to Schumpeter, the assumption is that 'the people' hold a definite and rational opinion about every individual question and that they express these opinions by choosing representatives who will try to carry out those opinions.[15] Schumpeter finds that this assumption makes it possible to assume that democracy is first of all concerned with ruling and not with the task of election, but he finds this basic assumption to be false.

Schumpeter conceives the election of the government to be the primary concern of a democracy. He takes the view that the role of the people is to produce a government (or an intermediate body which could produce a government), and he says: "The democratic method is that institutional arrangement for arriving at political decisions in which individuals acquire the power to decide by means of a competitive struggle for the people's vote."[16] Democracy then becomes a formal characteristic; it does not have to do with the actual questions to be dealt with in governing. Schumpeter underlines the fact that in a democracy the primary function of the elector's vote is to produce a government. In fact Schumpeter maintains that producing a government practically amounts to deciding "who the leading man shall be".[17]

Schumpeter's interpretation is provocatively simple, but let me point out one of its 'advantages': It has great descriptive value. It explicates how the concept 'democracy' normally is used when countries describe themselves as being democratic. Schumpeter underlines that his interpretation is quite close to reality in the case of the U.S.A. Besides, if we accept Schumpeter's definition, it becomes superfluous to think of democracy as an educational concern. Critical citizenship becomes a hybrid construction, and Dewey's formulations becomes devalued as empty rhetoric.

This interpretation dashes all idealistic dreams about direct democracy as well as (naive) expectation about the rational behaviour of humanity. In addition, it neglects every concern about the non-formal conditions for democracy, such as the fair distribution of goods and equal opportunity. Even the basic idea that democracy should be a way of keeping power in the hands of the people is eliminated. It is also obvious that if we take Schumpeter's interpretation literally, we need not care any more about the existence and nature of democratic competence. Democracy does not presuppose participation and has nothing to do with decision-making, or with criticism and evaluation of decisions and proposals put forward by the government. Democracy has to do only with production of the government. It consists in the procedures and algorithms for election. This is the most simple and radical solution to the problem of the delegation of sovereignty. The only input from the people to democratic life is the vote. Apart from that, they have only to receive.

The concept of democracy to which I subscribe does not have the great descriptive value of Schumpeter's. It is much closer to the classical interpretation which regards democracy as a characteristic of governing, although the form of the actual production of a government should not be ignored. I shall not return to the utopia of direct democracy, so the delegation of sovereignty has to be dealt with. Democracy also characterises the ways of participating in discussions and criticism of the actual ruling process. A democracy must make room for a critical citizenship, which is the actual display of a critical competence. So the problem of democratic competence is back on the agenda. If everybody behaves rationally and if rationality is sufficient, the display of a democratic competence is not going to be any problem. However, I do not rely on the rationalistic optimism that the only thing needed for normative judgements is reason. I find Schumpeter right in his criticism of the assumption of the actual rationality of humanity. But the answer is not to dilute democracy into procedures for producing a government. We shall address the problem about the conditions for developing a democratic competence directly. This competence is not born naturally. It does not grow from rationality. It has to be established and developed. I see this as one of the fundamental conditions for democratic life. We have, therefore, to discuss the possible content of democratic competence. It must be characterised in relation to the main questions which concern the society in question. The content of the democratic competence depends on the nature of the problems which face the society. And what are the conditions for establishing a critical citizenship in a highly technological society?

4. A PROBLEM OF DEMOCRACY IN A HIGHLY TECHNOLOGICAL SOCIETY

Democracy may be destroyed by a dictatorship which obstructs formal democratic procedures. This has been seen often and is normally conceived as being *the* problem of democracy. It is what some countries accuse some other countries of ignoring. But unless we accept Schumpeter's interpretation, the algorithm for the election of the people in charge is only one aspect of democracy. Democracy can be undermined in ways other than by just neglecting rules of election. Democracy refers not only to formal, but also to material and ethical conditions and to possibilities for participation and reaction. In particular, democracy can be destroyed if a critical citizenship cannot be brought into life. A dictatorship is not the only danger, also 'distance' has a role to play. If the distance between those in charge and the people is too great, then it becomes impossible to establish a relationship having to do with the content of democracy.

When society has grown to a certain level of complexity, the main principles of the mechanisms of the development of society become hidden and difficult to identify. How is it possible to evaluate decisions taken by the people in charge, if neither the conditions for, nor the implications of, their decisions are detectable? How can anyone other than experts control experts? Will the conditions for critical citizenship be eroded by social and technological development itself and the consequence become an 'expertocracy'?

This problem can be discussed in relation to observations made in an article by Daniel Bell, which has become a classic, 'The Social Framework of the Information Society'. According to Bell, the information society is characterised by the 'codification of theoretical knowledge' as its axial principle. This is underlined by a new theory of value. Bell finds that knowledge, and not only capital and labour (as specified in economical theory by the Function of Production) are the basic sources of value, because knowledge becomes involved in a systematic form in the transformation of resources. In the information society the ability to collect, systematise and use information becomes the vehicle for social development. Simultaneously, it becomes a source of power.

This raises the question of power structures in the information society (or the highly technological society, or the post-industrial society). A new category of dictatorship may emerge. The question is whether an 'expertocracy' becomes a threat to democracy in a highly technological society. It is interesting to consider some of Bell's comments: "In the post-industrial society, the technical elite is a knowledge elite. Such an elite has power within intellectual institutions – research organisations, hospital complexes, universities and the like – but only influence in the larger world in which policy is made. Inasmuch as political questions become more and more intricately meshed with technical issues (from military technology to economic policy), the knowledge elites can define the problems, initiate new questions, and provide the technical bases for answers; but they do not have the power to say yes and no. That is a political power that belongs, inevitably, to the politician rather than to the scientist or economist. In this sense, the idea that the knowledge elite will become a new power elite seems to me to be exaggerated."[18] A little later Bell concludes: "The fear that a knowledge elite could become the technocratic rulers of the society is quite far-fetched and expresses more an ideological thrust by radical groups against the growing influence of technical personnel in decision-making."[19] Bell does not see any danger in an expertocracy emerging.

I shall accept Bell's indication to some extent. I shall not look upon expertocracy as a real danger, as far as this is interpreted in a personal form. However, this does not solve our problem about the existence of democratic life. Let us assume that Bell is right: the elite with expertise may have power within intellectual institutions but only influence, not real power, in politics. Decisions are made by political institutions. But the problem of democracy, raised by the development of a highly technological society, is not just that of the influence or power of a knowledge elite placed outside the arena of political decisions. The problem is not just whether or not technological development reduces politicians to marionettes giving voice to the consequences of prefabricated technological calculations. The problem concerns the relationship between, on the one hand, the people in charge (the elected politicians) and the technological elite and, on the other hand, the people who are affected by the ruling. So even if the knowledge elite is only able to exert influence while the politicians remain in power, the conditions and the reasons for the decisions to be made may go beyond the reach of ordinary people.[20]

According to Schumpeter, we do not face any problem of democracy, even if a knowledge elite develops the conditions and the arguments for the decisions to be taken by the politicians. When democracy has to do only with the production of a government, the problem of the existence of democratic life completely disappears. However, I shall refuse to rely on this 'solution'. Therefore, we have to return to the problem that the ground for decisions taken by the authorities may be inaccessible to people other than the technicians and the people in charge. Technological development may *erode* part of the non-formal conditions for democracy, leaving behind only an algorithm for election.

That erosion is a real threat to democracy in a highly technological society. But is it possible to secure a critical citizenship in a highly technological society? To find a positive answer to this question is equivalent to conceiving democratic life as possible, in the future as well as the present. We shall not try to turn the clock back. We cannot propose abandoning a technological environment. Therefore, we cannot expect the conditions for a critical citizenship to decline. The problem is to develop a critical competence which can match the actual social and technological development. This constitutes one of the essential *problems of democracy in a highly technological society*.[21]

5. EDUCATION FOR *MÜNDIGKEIT*

Which institutions in society can take over the job of developing a democratic competence? It cannot be assumed that this can be done in any straightforward way, because the content of such a competence will change in accordance with the development of society. It is a socially defined competence. One answer given is that education must be in charge.[22] Traditionally a major concern of education has been to prepare students for later participation in the economic processes of society, as education must distribute the skills necessary for society to function. But, as we have seen, different trends in education have stressed that education must also prepare individuals to deal with aspects of social life outside the sphere of work. To pursue such aims has been a strong trend in German education, especially since World War II. It is underlined by the use of the term *Allgemeinbildung* (general education or liberal education[23]), meaning that education must focus outside the conditions for generating possibilities for work. In the Scandinavian tradition also, the idea of a general education has played an important role. Education must prepare students for (political) life in society. This is what both Dewey and Adorno have emphasised, and the educational concern for democracy and critical education come to share the task of empowerment for a critical citizenship.

In critical education the term *Allgemeinbildung* is categorised into different, more 'active' operations: education must assume an active role in identifying inequalities in society, in identifying causes for the emergent sociological and ecological crises and in explaining and outlining ways of dealing with such problems. What has been emphasised up to now, however, is perhaps the re-active part of a democratic competence, mainly having to do with criticism and evaluation of suggestions or actual decisions to be taken by the people in charge. The active part of being a citizen has not been particularly stressed. But from the perspective of critical education this part is important also. The utopian aim is to be on equal terms with the authorities. This is what is entailed in the previously quoted formulation of Giroux stating that schools must be defended as an important public service that educates students to be critical citizens who can think, challenge, take risks, and believe that their actions will make a difference in the larger society.

It may be helpful to take a closer look at one of the concepts used by Adorno (as well as by other German educators). 'Education after Auschwitz' was published in the book *Erziehung zur Mündigkeit*, and here *Mündigkeit* needs explaining. The concept has a double meaning and unites perspectives developed in relation to both 'critique' and 'democracy'. The term relates to legal theory referring to a person obtaining full age and therefore acquiring an adult's rights and responsibilities in society. The person of full age has the legal right to speak for himself or herself, and to vote, for example. But the term also has an informal meaning: that of having the capacity to speak for oneself. In that way *Mündigkeit* becomes an essential constituent of a critical citizenship. It unites features of a democratic competence together with a critical capacity. A person with *Mündigkeit* shows the capacity to take well-balanced decisions. Therefore it makes sense to educate for *Mündigkeit*, not to educate 'followers'. The main task of education is to prevent the occurrence of a new Auschwitz.

The concept of *Mündigkeit* is embedded in the discussion of literacy, as expounded by Giroux, and it raises again the question: How does the fragmentation of the curriculum into different disciplines affect the conditions for the development of critical citizenship? The position of a specific school subject may play an important role in this context, namely mathematics. Do literacy and mathemacy have similar roles to play in an education for *Mündigkeit* in a highly technological society?

CHAPTER 3

MATHEMATICS – A FORMATTING POWER?

Nicolaus Copernicus created a new world in the sense that he changed the fundamental interpretation of humankind's position in the universe. Charles Darwin also changed the status of humankind by connecting it closely to (other) animals. These points demonstrate that scientific knowledge can change our basic understanding of the world. Science has a strong interpretative power. However, scientific development can change not only interpretations of reality, but reality itself. Physics and chemistry both have changed our understanding of nature and, in addition, the object of our understanding, i.e. nature itself.

That empirical sciences can cause such change is perhaps not surprising because these sciences have 'reality' as their object. But what about mathematics? Normally, mathematics is conceived of not as an empirical but as a formal or abstract subject, and therefore a mathematical proof cannot give evidence of any real phenomenon, only evidence for abstract mathematical statements. According to Platonism the world of mathematics exists entirely apart from that of the empirical world, and mathematical propositions deal just with pure 'forms'. Formalism refrains from creating a Platonic reality but does separate mathematics totally from the empirical world by interpreting the subject as the science of formal systems. The neo-empiristic interpretation of mathematics does not find this separation to be a fundamental one. However, the discussion about whether neo-empiricism provides a more adequate philosophy of mathematics or not is not essential in the present context, because I want only to highlight the question: Could a science like mathematics (formal or not) become not only interpretative but also formative? I want to discuss this question independent of the controversy between on the one hand Platonism, formalism and other variants of 'apriorism' and on the other hand, the neo-empiricist interpretation of mathematics.

The thesis I want to put forward is that mathematics produces new inventions in reality, not only in the sense that new insights may change interpretations, but also in the sense that mathematics colonises part of reality and reorders it – just as the Indian social reality was changed after the English conquered the country and took power. This idea contradicts, for instance, G. H. Hardy when, in *A Mathematician's Apology*, he stresses that real mathematics is of no use and possesses a sublime pureness. If mathematics is able to colonise, the subject has completely lost its innocence. But how can this be possible? The thesis is that *mathematics is formatting our society*. However, I do not wish to argue this thesis, my intention is only to give it more meaning.[1] The idea that mathematics has a formatting power is a specification of the thesis of linguistic relativism and at the same time a generalisation of this idea, because the distortion of a perspective, caused by its being situated in a given language reality, may in fact more easily become 'objectivised' when it is situated in a reality of mathematics.

It may be useful to mention what the thesis does not mean. First, it does not say that mathematics is the only agent in performing social formatting. The thesis is about an interplay between agents within social development, and mathematics is suggested to be one of these agents. The actual role of mathematics, the way mathematics may interact with other social forces, has therefore to be expounded. Second, the thesis does not say that mathematics itself is immune from social determination. Even if mathematics can be seen as a social force, it can still be conceived as a social construct. Third, naturally an abstract subject cannot be doing anything by itself, the thesis needs a sociological interpretation, and the interpretation is not, for instance, that mathematicians as a social group are assigned any special importance. The thesis concerns the outcome of a mathematical research paradigm. So, perhaps God did not organise the world according to mathematics, but the thesis suggests that humanity has now embarked on just such a project.

1. TECHNOLOGY AND THE VICO PARADOX

Applications of mathematics have shown growth of epidemic proportion. Previously, subjects like physics, astronomy and chemistry were the main areas for applications of mathematics, but now no subject area seems to be immune from quantitative analysis. The escalation of the applications of mathematics is closely connected to the development of information technology, making it possible to handle complicated mathematical models and overwhelming quantities of data. Information technology may be seen as the materialisation of mathematics, or we may conceive mathematics as an entity hidden or 'frozen' in the computer. However, before returning to this idea, I shall make some comments on technology in general.

Technology is often interpreted as a basic condition for social development. This statement can be turned into a deterministic attitude: Technology is seen as the dominant factor and therefore technological progress produces a social organisation. If this deterministic attitude is connected to a pessimistic view of technology, fatalism will be the consequence. The main trend in technology is not seen as approaching a solution to (technological) problems but, primarily, as creating problems itself – problems which may not be soluble by any technological innovation. Determinism can be connected to an optimistic view as well: We will always be able to solve our problems by new technological improvements, as the technological development itself is oriented towards an attractive future. It is, however, possible to leave the deterministic position and to maintain the existence of a space of freedom in technological development. For instance, it is possible to maintain that, although technology has no immanent direction, human endeavours can be so directed that, because of the rich potential of technology, social progression can be diverted towards the most attractive direction. Technology as such has no tendency towards 'good' or 'bad', but potentially it contains solutions to human problems. More generally: Technology comes to reflect 'good' or 'bad' features of social and economic structures.

In the philosophy of technology ideas about liberation figure side by side with fears of suppression, controlling and disciplining. We can see liberation as liberation from the straitjacket of technology, as a liberation from the power and command by which 'the technological necessity' confronts us, and as liberation from the pitfalls which technological development continuously is digging. However, technology has been linked with an interest in liberty as well, and from this outlook technological development provides the fundamental conditions for realising some of the attractive potentials of human life. Originally an optimistic interpretation dominated the philosophy of technology, while the pessimistic trend arose later. This means we witness a shift from seeing technology primarily as a means for critique, in the sense that a technology may help us to tackle a critical situation, to seeing technology as a possible object for critique, being in itself the cause of critical situations.

Liberation can be expounded in two different ways. It is possible to talk about freedom *from* – and then to mention poverty, deprivation, hunger, etc.; but we can also talk about freedom *to* – and mention freedom to think, to discuss, to act, etc., whatever the individual prefers. The different contributions to the discussion of liberty have not emphasised these two variants to the same degree. From a materialistic perspective, freedom *from* has priority because this form of liberty is supposed to provide the conditions for further liberation. Opposing this, liberalism (and anarchism) see personal freedom as the basic condition for human life and therefore as a necessary condition for further emancipation. In the philosophy of technology both interpretations of liberation have been mentioned, for instance, when Friedrich Dessauer, in his book *Streit um die Technik*, emphasises that technology means freedom in a double sense.[2]

The conception of technology as a means for extending human freedom has normally decoded technology as a tool which humanity inserts between itself and nature. Technology is viewed as a human-nature relationship, and technology provides human beings with power to cope with nature (originally looked upon as a hostile condition for life). In *Grundlinien einer Philosophie der Technik*, Ernst Kapp explains technology as an extension, or amplification, of human organs. The hammer is an extension of the fist and the arm, etc. In this way technology becomes a tool which human beings can use in their struggle for survival. This interpretation is rather common. As an example let me refer to *Einführung in die Philosophie der Technik* in which Heinrich Stork, as an initial definition, describes 'technique' as an action by which humanity consciously changes natural structures and energy in such a way that they can serve the needs and uses of humankind.[3] Many philosophers of technology continue to see technology as an 'instrument' by which humanity tries to cope with nature, and, as a consequence, technology is conceived of as a means of liberation by overcoming the limitations which nature imposes on human life.[4]

In *The Technological Society*, Jacques Ellul extends the perspective of technology, as does Herbert Marcuse in *One-Dimensional Man*. For both authors, a cardinal point is that technology has a profound influence on society. Therefore it only becomes possible to discuss the function of technology by using a broad interpretation based on perspectives from social sciences and a variety of disciplines from the humanities. The reason is that these authors see technology as a fundamental structuring principle. Technology concerns all aspects of social life, it becomes an absolute for social organisation. Not only the form of production but the whole 'civilisation' undergoes a technological reconstruction. Nature in its original sense disappears, and we become inhabitants in a technological reconstruction of our social reality.[5]

Both Ellul and Marcuse find a decisive tendency in the submission of humanity to technology, and that means that technology becomes located in the opposite direction from freedom. It becomes related to suppression, domination and disciplining. The one-dimensional human being is kept in check and placed under tutelage. On this point, however, we see a diversity between Ellul and Marcuse. While Ellul outlines a pessimistic view – he sees society and the social development as entirely determined by the technological development itself – Marcuse maintains the possibility of an alternative technology, and this makes it possible for him, in the midst of his pessimism, to establish a belief in the possibility of influencing technological development. Therefore the connection between technology and subordination is not a necessary one; it is still possible, in utopia, to relate technology to an idea of freedom.

Following Ellul and Marcuse, we see not only a change by broadening the discussion of technology, we see also a change in the concept of technology. Initially, technology was conceived as a relationship between humanity and nature, but in the hands of Marcuse and Ellul technology also becomes a relationship between human beings. The object of technology need not be nature but can be humanity and social structures. (Even if technology can be used in the struggle against nature, it seems always to be the case that the same technologies can be used in warfare. This highlights the observation that there is "no document of civilisation which is not at the same time a document of barbarism".)[6] This observation can be extended, and I shall emphasise the existence of different types of technology: tools, energy technology, social technology and information technology. We have here a division in types and not of categories. I do not maintain that the suggested number of types is exhaustive, for instance, it could be claimed that biotechnology constitutes a type of its own (it may also be maintained that this technology primarily relates to information technology); however, this is not essential to what I want to say.[7] The division into types makes it possible to elaborate a bit more about mathematics and technology.

Tools can be seen as a generic technology. As mentioned, Kapp finds tools to be paradigmatic for technology and therefore his philosophy of technology develops as generalisations arising from interpretations of tools defining a relationship between human beings and nature. By contrast to other types of technology, a tool can be characterised by its 'integrity'. The user of the tool has to provide his or her own energy or to add to this energy from 'natural' sources, like wind, water, draught animals, etc. The element of energy as well as the element of information remain implicit in the working situation when we consider tools.

The object of energy technology is nature as well, but now nature becomes interpreted as a physical system. Natural power and muscular strength are substituted by machine power, and the new source of energy makes us able to produce enormous 'tools', and in doing so we step into the age of industrialisation. Energy technology does not replace tools, it substitutes natural power, and by so doing new constructions of machinery become possible. Marx sees energy technology as constituting a philosophy of technology, and so the analysis of the ownership of the means of production becomes an essential factor in his economic theory. Because of energy technology the 'machinery' of industrialisation grows into such a massiveness that it becomes immobile, and therefore it becomes possible to possess technology in the same sense as one can own land. When the means of production are 'fixed' and owned, new conditions for production are established, for instance, the workers have to come to the factory.

Social technology does not have nature as its object but is directed towards relationships between human beings. I see 'scientific management', as developed by F. W. Taylor, as a paradigmatic example of this type of technology. The basic idea in scientific management is that complex work-processes have to be fragmented into their atomic components. Then each component can be investigated in order to find out the best way to handle the operation, and the 'proper time' for its execution can be measured. Then the atomic components have to be sequenced to define the complete work-process for each worker, and the total addition of the workers' algorithmic behaviour will make up a new 'mega-machine'. Taylor described how he investigated specific work-processes to identify the best way for their execution and, as an example, he gives the story of the 'perfect' worker, named Schmidt, who never asked questions, but just followed the prescribed algorithms. Schmidt thus became the first Taylorised person.

The last type of technology mentioned is information technology. This technology can also be seen as a sort of substitution – not muscular power, not rhythms of work but 'brain power' is now available as a substitute. I choose not to take into consideration a variety of traditional forms of information technology, like printing, telegraphy, broadcasting, etc., but choose instead to concentrate on computer-based information technology. This does not impose any vital limitation on what I am going to say about mathematics and technology.

We can try to summarise by mentioning an exemplary piece of 'machinery' relating to each type of technology, but it must be emphasised that I do not identify technology only with its physical components. In my conception technology refers to machinery as well as to a competence in constructing and applying it. Nevertheless, the hammer can be seen as the exemplar of tools, the steam engine as the exemplar of energy technology, and, as we have limited the interpretation of information technology, the computer becomes the exemplar of that technology. Finally, we have social technology, and I find the watch to be the exemplar of this technology.

This short foray into interpretations of technology – being optimistic or pessimistic, deterministic or open-ended – shows the uncertainty about how we can look at the phenomenon. What displays the difficulties in coming to grips with the basic features of a technological society especially is the change of focus from seeing technology as a human-nature relationship to seeing it as a human–human relationship, specified into different types of technology. This difficulty can be summarised as the *Vico paradox*.[8] Giambatista Vico was an Italian philosopher who formulated the idea that the only things possible for human beings to understand are what they have created themselves. This represents an attack on Cartesianism, which turns towards Nature as the object of infallible knowledge. Vico expressed a reservation about the ability of human beings to grasp nature. He doubted that we should ever come to understand any creation of God, we have the capacity to comprehend only our own creations.

Following this line of thought we should be able to understand technology, which is *per se* a human construction. 'Understanding by doing' seems to be a most reliable epistemological thesis. However, we seem completely unable to establish such an understanding in the case of technology. We seem to be without the capacity for grasping the limits and the full consequences of our technological enterprises. This statement reiterates what has been indicated by the thesis of linguistic relativism: Being submerged in language, we have difficulty in comprehending the function of the linguistic structure itself. And in being submerged in technology, the situation is similar. Although technology is a human construction, we do not seem to posses the capacity to comprehend what we have constructed. This I call the Vico paradox. In fact this paradox is just a different way of stating the problem of democracy in a highly technological society.

2. MATHEMATICS AND TECHNOLOGY

Differences between types of technologies also become visible when we concentrate on their relationships to scientific parent disciplines. With regard to tools this relationship is accidental. Illustrative of the fact is that in ancient times we do not find any intimate connection between the development of tools and natural sciences. In the development of tools scientific knowledge did not play an essential role. Following the ethnomathematical approach, we can state that implicit in the constructions and applications of tools we may find mathematics, but mathematics remains a tacit competency which it is not necessary to explicate in order to improve the intended applications of the tool. Naturally, it is possible to find exceptions, for instance, mathematics figures as an essential condition for the construction of maps (which I interpret as a sort of tool).

The situation is different in relation to energy technology. Although many energy technological innovations do not build directly on the natural sciences or mathematics but rather on previous technological experiences (we have only to think of the development of the steam engine) we also find a sort of 'parallelism', because the principal understanding of the possibilities and limitations of energy technology are obtained only in the natural sciences. When we come to the development of the frontiers of energy technology, we find that the natural sciences become the substantial basis for technological innovations.

Social technology, however, does not rest firmly upon any scientific discipline but is linked to economy, sociology and psychology in an *ad hoc* way. It may be maintained that social technology implicitly builds upon a paradigm borrowed from the natural sciences – thus social technology often interprets its subjects, human beings and social relationships, from the perspective of the strategic attitude of how they can be manipulated. Mathematics has a solid position in social technologies, but in this case the position of mathematics is often different from its integration in the natural sciences. Frequently mathematics simply facilitates *ad hoc* techniques.

Information technology has a quite different, intimate relationship with science, because the construction of the computer is closely related to mathematics and logic. Mathematics and related subjects provide conditions for the development of information technology, and it is distinctive that the nature of the computer and the possible limitations of algorithmic procedures were discussed before the construction of the first computer. In fact, every application of a computer can be seen as an application of a simple or complex mathematical model, and because information technology appears to be omnivorous, mathematics moves into a new social position. Therefore, from a logical point of view, information technology need not be interpreted as a new way of formal manipulation. An effect of computing is that applications of formal methods come to colonise all areas of life. Simultaneously, information technology has a special position in the family of technologies in that other types of technologies become dependent on this technology, if not absorbed by it.

It is maintained that a dominant type of technology determines the fundamental structuring principles in society, and following this idea it has been emphasised that information technology provides a new condition for social life. A strong thesis is put forward by Bell who, in the article mentioned previously 'The Social Framework of the Information Society', states: "In the coming century, the emergence of a new social framework based on telecommunications may be decisive for the way in which economic and social exchanges are conducted, the way knowledge is created and retrieved, and the character of the occupations and work in which men engage. This revolution in the organisation and processing of information and knowledge, in which the computer plays a central role, has as its context the development of what I have called the post-industrial society."[9] In this general formulation we find the assumptions firstly that technological development may provide new social structures and secondly that the highly-technological society is characterised by the dominance of computer-based information technology. Therefore, the conclusion seems to be that mathematics may contain an overwhelming social power, being the principal science in that technology. This makes up the central claim of the thesis that mathematics is formatting our society. But still this seems to contradict the fact that mathematics is abstract and therefore apparently unable to 'touch' reality.

If we take a pessimistic interpretation of technology, we see some of the structural problems of society as critical and as caused by the technological development. We are simply unable to understand basic contradictions in society without understanding how they are created or influenced by technology itself. We especially cannot understand the structures of risk which technological development carries us towards without understanding the nature of technology. To carry out a critical analysis of our social situation therefore presupposes a conceptual framework by means of which we are able to grasp the formatting powers of society. Further, conflicts and crises have to do with the distribution of power and control, and that also becomes a way of drawing our attention to the problem of democracy in a highly technological society. The Vico paradox has not come closer to its solution because mathematics has obtained a central role in the development of technology. The transparency of mathematics has not produced a transparency of the actual functions of mathematical formattings.

3. ABSTRACTIONS

In what way can a science dealing with abstract objects be able to do formatting? We shall distinguish between two different kinds of abstraction: *thinking abstraction* and *realised abstraction*.[10] Thinking abstraction refers to the mode of thought used to facilitate reasoning, and this type of abstraction is exemplified by mathematical concepts and mathematical modelling. Here we are concerned with explicit mathematics. In trying to launch a satellite we may use a model of the earth, abstracting from the roughness of its surface, and we may try to calculate a possible orbit of the satellite. Instead of making a real experiment we are able to deal with the situation hypothetically. Thinking abstractions can facilitate thought experiments. Reasoning about economic development may be helped by using a concept like National Product, defined as a mathematical function, the value of which is determined by the values of different parameters.[11] A variety of assumptions and also misunderstandings may be incorporated in a thinking abstraction. Thinking abstractions exist only as mental models or images. Their existences are similar to that of a character in a novel.

This can also be expressed in terms developed by Karl Popper making it significant to interrelate three different 'worlds'. According to Popper, the first world consists of physical entities, i.e. the world of facts and state of affairs, the second world is made up by the individual subject's private state of mind, while the third world consists of scientific ideas. The most important inhabitants of this third world are scientific concepts and theories. An essential question concerns the relationship between the three worlds, and Popper maintains that – although he provides the third world with an independent objectivity and existence – it is possible to maintain consistently that its entities are the result of human production. The third world is an objective human construction, and the logic of scientific discovery, which Popper specifies as an alternation between conjectures and refutations (explained by Imre Lakatos in the case of mathematics as a dialectics between proofs and refutations), becomes the form of change in this world.

What we are going to consider is whether it is possible to find a link between the third world (of thinking abstractions) to the first world even in the case of a science like mathematics. Let me, however, make a substantial addition to the third world. While Popper conceives this world as consisting of scientific concepts and theories, I shall include generally accepted assumptions and ideas as well, even if they do not constitute well established knowledge. Nevertheless, they can function as modes of thought. This addition is important in our context, for although a mathematical concept as such may be conceived of as scientific, a part of mathematics that has been interpreted and used as a model need not be scientific-based and can in fact express a biased opinion as well as different misunderstandings. In short, the interpreted part of mathematics figures as a thinking abstraction. To give a short illustration:

The Ptolemaic interpretation of the positions of the sun and the planets is an interpreted mathematical model which belongs to our thinking and which creates ways of looking at reality.

Realised abstractions have a different ontological status than thinking abstractions. Thinking abstractions are (though perhaps rather imprecise) 'images' of reality, but we also may witness the reverse phenomenon that real structures can be 'images' of thinking abstractions, and these we call realised abstractions. They are taken for granted and become reifications of modes of thought. Normally we do not question whether what we are dealing with is a realised abstraction or not, it is real and belongs to our first world, to use Popper's terminology. Perhaps we do not even have the possibility of identifying a realised abstraction. We live together with such abstractions. Ways of calculating taxes are not just thinking models, they have a real influence on our lives. Exchange values of goods in the form of money are real, they are not just models for expressing the degree of usefulness of some goods, or for expressing the time necessary for their production. Money becomes a real state of affairs, and even the National Product becomes real: it reaches a status different from being just the mathematical summary of calculations based on specific values of certain parameters. The National Product enters the political and economical discussion as an independent object and a real figure.

Every society and every culture has developed a realm of realised abstractions. But from what sources? They must be brought into existence by some creative act. We may, for instance, be able to trace some realised abstractions back to ideological structures or to metaphysical systems. However, as realised abstractions they have obtained the status of laws and principles for the formation of certain social entities. They have to be taken into consideration as part of our reality. They are not any longer just models for our thinking. So, while a character in a novel remains in the novel, some thinking abstractions change their ontological status and they materialise. The concepts come to life. As mentioned, it may be possible to trace the origin of differently realised abstractions back in time, but what is important in this context is that some realised abstractions in a highly-technological society may emerge from mathematical modelling. The assumption is that this subject becomes a new, bubbling source for the invention of rules and structures, i.e. mathematics not only creates ways of describing and handling problems, it also becomes a main source for the reconstruction of reality.[12]

4. FORMALISATIONS

We shall try to get a better understanding of how thinking abstractions may become realised abstractions by looking at different kinds of *formalisation*. A distinctive example of axiomatisation, the first important step of a formalisation, is given by David Hilbert. He describes Euclidian geometry by laying down axioms connecting the concepts of 'point', 'line', and 'plane'. No explicit definitions of these concepts are given, only their relationships are stated. In this sense Hilbert uses implicit definitions as starting-points for his axiomatic system. But do 'points', 'lines', and 'planes' really exist? Euclid has already tried to answer this question by suggesting explicit definitions of these concepts, but he did so without success. Hilbert's criterion for mathematical existence is surprisingly simple: The objects, referred to by the fundamental terms in an axiomatic theory, exist if and only if the theory is consistent. This interpretation makes mathematical existence abstract, which is also underlined by the fact that Hilbert found that 'ideal' objects can be added to a theory to put a finishing touch to it – of course the condition still is that the theory remains consistent. When a mathematical theory is axiomatised the problem of existence is, as it were, extracted from philosophy. Existence becomes a formal property and the content of mathematics disappears from our world. When a theory is axiomatised, it can be formalised, and by so doing we *formalise a language* or a part of a language. In this sense formalisation is an act by which a new language game is created.

When we have formalised a part of a language we have created a thinking abstraction, but naturally other ways of producing thinking abstractions also exist. A language provides not only a description of reality, it also provides an interpretation. Depending on our means of description we are able to identify different aspects of reality. In fact, if language provides structures of an interpretation from which we are not able to escape, it does not make sense to talk about reality in isolation from any interpretation. We are not able to put language outside the parenthesis of our consideration. Instead, we are absorbed in our language and by our language. We can try to change the language in use, to elaborate new means of expression, but we cannot do this from outside language. Some of the possible languages available are the formalised ones. To formalise a language and apply it therefore means to create a new interpretation.

Formalism gives the most precise description of a formal language, and by so doing it sets up an ideal for the formulation of a mathematical theory.[13] We shall restrict ourselves to taking a short look at the formalistic definition of a formal language. A formal language rests upon an unambiguous grammar. It has to be well-defined whether a given symbol belongs to the vocabulary of the language or not, and it has to be well-defined also whether a sequence of symbols makes up a well-formed formula or not. The grammar has to generate exactly the class of meaningful sentences of the language. When a number of well-formed sentences are stipulated as axioms and a system of deduction is specified, we can develop the formal theory (and also try to investigate whether the system is consistent and complete or not). When we have formalised a part of a language we have arrived at a thinking abstract whose ontological status is indicated by Hilbert's interpretations of mathematical existence. It belongs to our thinking only.

We have to take a look at another type of formalisation. It is possible not only to formalise language, but also to *formalise actions and routines*; i.e. ways of behaviour. In this case the result of a process of formalisation is not a new language game but a new way of behaving, including 'manuals', i.e. descriptions of how to behave in a predescribed algorithmic way. This phenomenon may become illustrated by 'scientific management', in fact I conceive Taylorising to be a paradigmatic example of the formalisation of actions and routines. When structures are set up, norms for behaviour are created simultaneously. We can think of planning a certain production: The workers have to behave in a specific way, and at the extreme end of the spectrum we find the fully automatised factory.

Formalisation of language and formalisation of action are closely connected because it is possible to move on from a formalisation of language into a formalisation of routines. The application of a formalised language in order to describe something real makes it easier to 'see' structures and, by doing so, a first step is taken in adapting reality to our image of reality. A description highlights some aspects and ignores others, as already noted: A description provides an interpretation. If, for instance, the object for our description is work-processes or economic transformations and our intention is to do further systematisation and Taylorising, a formal description will facilitate our steps. It makes it easier to identify new algorithms of behaviour. This exemplifies the point that our reality will always be our interpretation of reality, and that language becomes projected into reality, including formalised languages.

What we witness is the passage of thinking abstractions into realised abstractions, caused by the transformation of a formalisation of language into a formalisation of actions. Abstractions become realised and, therefore, we may also talk about *real abstractions*. We create a semantic for our formal description by inventing procedures, algorithms, routines, i.e. sorts of behaviour referred to by the formal language. Mathematics intervenes in reality by creating a "second nature' around us, by giving not only descriptions of phenomena, but also by giving models for changed behaviour. We not only 'see' according to mathematics, we also 'do' according to mathematics. In this sense inhabitants from Popper's third world materialise in the first world.

Although mathematics may be interpreted as a formal subject and mathematical existence as being separated from real and empirical existence, it is possible to see an ontological transformation of formal structures into empirical and social realities. Mathematical models become guidelines for our design of our world and, therefore, they become not only descriptive but also prescriptive. Because we base the design of our social environment on mathematics, we come to live in a world of 'second nature', inhabited by strange new formal creatures, which constitute an integrated part of society. They cannot be substituted by any other abstractions serving similar functions. It is impossible to imagine the development of a society of the type we know and live in without realised abstractions manifested via information technology.

Mathematics *constitutes* realised abstraction in the same way it constitutes modern warfare. What does this mean? Not that warfare would not exist without mathematics, but to 'subtract' mathematics from modern warfare would mean the complete dissolution of that type of warfare. Similarly, we cannot imagine how a modern tax system could exist if mathematics were 'subtracted'. Naturally, these formulations are misleading, too, because the metaphor 'subtraction', so to say, presupposes a sort of Platonism: that mathematics exists and can be applied and that it can be 'subtracted'. The point is that mathematics is integrated in the technological structures.

A main means for realising abstractions is through 'systems development', which acts as a midwife in helping formal creatures emerge in life.[14] They become replicated. New generations grow up with the speed of germs. A technological world is created around us in a way which makes sense to Ellul and his ideas, seeing technology not only as a structure of society but as a fundamental category embracing the whole of society. We live *in* technology, *for* technology and *by means of* technology. Schmidt is no longer the only Taylorised person.

5. MATHEMATICS AS CRITICAL

Both thinking abstractions and realised abstractions constitute objects for critique. To address thinking abstractions, means to criticise ways of interpretations, i.e. mental objects; and to address realised abstractions, means to criticise something 'material'. In particular, this means a critique of what mathematics is doing. The step from thinking abstractions to realised abstractions has been described as a formalisation, and the whole process of formalisation therefore becomes an object for critique. To direct this critique is important, if formalisation is to be analysed in the same way as technology in general, i.e. as an activity which simultaneously solves and creates critical features of society. In particular, it is my assumption that the fundamental structures of risk which are part of a highly technological society have to be understood as an outcome of a formalisation.[15]

To talk about realised abstractions is a different way of stating the point that linguistic structures manifest themselves as part of reality. This linguistic relativism, however, does not specify much about the concept of power. It may be useful to look at the transition from thinking abstractions to realised abstractions and the nature of formalisations, including the step from formalisations of language into formalisations of routines, as expressions of *symbolic power*. We need not look for power in the shape of some overt physical arrangement. Power can be expressed and exercised in symbolic form. Symbolic power presupposes that those dominated believe in the legitimacy of the power. This is an important aspect of the applications of mathematics. The results of mathematical calculations have to be believed in order to be 'materialised' as working routines. The symbolic power of mathematics is rooted in a widespread metaphysics concerning the reliability of mathematics and this metaphysics may be a part of the hidden curriculum of mathematics education.

Pierre Bourdieu has discussed examples of symbolic power, including symbolic power in relation to education.[16] He emphasises how it is that people speaking local dialects are induced to collaborate in the destruction of their own means of expression; in this case education adopts the aim of imposing a new language, the 'official dialect', as the standard of education. This can become generalised, and we can see the concern of ethnomathematics as a resistance to the symbolic power of the 'official language' of mathematics. Bourdieu describes symbolic power as "a power of constituting the given through utterances, of making people see and believe, of confirming or transforming the vision of the world and, thereby, action on the world and thus the world itself, an almost magical power which enables one to obtain the equivalent of what is obtained through force (whether physical or economical)".[17] It is possible to interpret this as a description of power exercised by formalisation.

The potential of symbolic power makes it necessary to relate the discussion to democracy also. If a society is based on the use of manual tools, an idealistic interpretation of democratic competence remains plausible; no specific technological knowledge seems needed to evaluate the acts and decisions of the people in charge. Quite the contrary seems to be the case in a highly technological society. The content of democratic competence seems rapidly to acquire a tremendous complexity. On the face of it, only a limited group of people seem to be able to manage this complexity. In fact this competence seems to presuppose a certain amount of technological knowledge. How could anybody evaluate decisions which must take into consideration consequences of technological enterprises without a fair amount of technological knowledge? And in the information society, as described by Daniel Bell, this technological knowledge is based on mathematics. Our analysis of the problem of democracy seems to imply that technological knowledge has to be developed at all levels in the educational system, and further that we have to increase mathematical education as an integrated part of technology. It seems tempting to assume that in a highly technological society mathematical competence constitutes a major part of a democratic competence. However, this conclusion is *not* the point, although special attention must be paid to mathematics.[18] This has been pointed out by Mogens Niss in the following way: "It is of democratic importance, to the individual as well as to society at large, that any citizen is provided with instruments for understanding the role of mathematics. Anyone not in possession of such instruments becomes a 'victim' of social processes in which mathematics is a component. So, the purpose of mathematics education should be to enable students to realise, understand, judge, utilise and also perform the application of mathematics in society, in particular to situations which are of significance to their private, social and professional life."[19]

This brings mathematics education into the focus of a critical education in a dramatic way. A brief look through traditional textbooks indicates that it is unlikely that painstaking absorption into the curriculum will enable students to realise, understand and judge applications of mathematics in society. However, Niss is searching for instruments for understanding the role of mathematics in society and these instruments need not be mathematical in themselves. Empowering of the 'victims' of a symbolic power need not imply a forced and extensive study of the grammar of the language of power, i.e. an expansion of the mathematical curriculum. A more radical change may take place. Although mathematics plays an important role in technologies which determine social development, an understanding of the role of mathematics and formal methods in society does not need to be found within mathematics. The missing competence need not be identical with some mathematical competence. Democratic competences concern evaluation and criticism of actions and proposals, and it is not obvious that an accumulation of mathematical knowledge would add up to some ethical competence.

Several authors have identified culture as an important point of orientation for mathematics education.[20] In a highly technological society, culture also becomes determined by that second nature which is created by materialised formalisations. To understand culture, therefore, also means to understand how such materialisations become part of everyday life. To make this out to be an educational task is optimistic – and also naive, bearing in mind the nature of the Vico paradox. Nevertheless, I see the task of critical mathematics education as being to establish such an understanding , and this brings together the notions of 'critique' and 'democracy' with the concept of *Mündigkeit*.

CHAPTER 4

A THEMATIC APPROACH IN MATHEMATICS EDUCATION

The three previous chapters could be summarised so: It is important to make education critical, if it is not to degenerate into a way of effectively socialising students into a technological society and at the same time annihilating the possibility of them developing a critical attitude towards precisely this society. The thesis about the formatting power of mathematics suggests that a formal science may turn an invention into a reality, and this is what brings mathematics education into focus. An important question becomes whether or not this education will be able to provide students with a competence fundamental to a critical citizenship. Can we offer an educational interpretation of 'mathemacy' which parallels that of 'literacy'? If this question is ignored, mathematics education easily becomes a means for increasing a symbolic power with important social consequences. (Still we have to remind ourselves that we cannot claim to have 'proved' any statement about the necessity of instituting critical mathematics education.)

Up to now I have focused discussion on the semantical links between concepts like 'critique' and 'democracy'. But we could make a fresh start by looking at perspectives comprised in the planning of classroom activities; and in my attempt to approach a philosophy of critical mathematics education I shall include different examples of educational practice. The function of these examples is not to provide a broader basis for an empirical investigation; in particular, the projects to be described are not set up in order to confirm some well-specified educational theses, nor are the examples offered as exemplifications of an educational philosophy. The teachers carrying out the projects did not try to 'follow' some explicitly formulated principles. Instead the examples serve as multifarious 'illustrations', in the sense in which a Pieter Breugel picture shows the richness and variety of life in a particular situation. The examples elucidate the complexity of educational practice and in this complexity I shall try to look for aspects of critical mathematics education. Such an approach not only informs us of the practice but is also a means for highlighting the concepts I use. This hopefully serves as a philosophical clarification, and a way to look for a meaning of critical mathematics education. (It is not the classroom situations as such which are the object of my interpretations, but the conceptions integrated in the planning of those activities. This does not imply that I find classroom observations unimportant, even in developing a philosophy, but simply that this has not been my concern.)

The 'pictures' of educational practice are not selected at random – not all Breugel pictures tell us about activities of children (but the pictures which do, also tell much more). The examples of mathematical project work are carried out by very competent teachers, interested in interdisciplinarity and in involving students as real participants in the educational process. At the same time the teachers are committed to a personal view of educational practice and to organising their teaching in a way they find appropriate to their overall experience. My job has been to interpret descriptions of their practice. (The descriptions are all based on the teachers reports and not on my observations.) In later chapters, I describe the projects "Golfparken", "Constructions", "Family Support in a Micro-Society", "Our Community" and "Energy". In this chapter, I shall describe an example of a thematic approach to mathematics education for 10–11 year old children.[1] The planning of this project, "Economic Relationships in the World of a Child", was a cooperative effort between the two teachers involved and myself, while the actual experimental teaching was done by the teachers themselves with their respective classes. The teachers involved were Marianne Klöcker and Inge Lise Kristoffersen from Egelundsskolen in Albertslund. The project took place in 1979.

1. SOME SOURCES OF INSPIRATION

In the mainstream of critical education, which addresses general questions of education, mathematics has normally been conceived of as a stiff and formal subject with a high resistance towards interdisciplinarity and cooperation, but also as a subject not essential to the establishment of a critical education. In the work of Negt, Mollenhauer, Lempert, Freire and Illeris mathematics is hardly mentioned.[2] These approaches do not raise the question whether 'mathemacy' and 'literacy' have similar roles to play.

The thematic approach I shall describe does not claim to exemplify critical mathematics education. The conceptual structure behind the planning process was rather loose and did not explicitly take into consideration those concepts which have been outlined in the first three chapters. The concept of critical education however was well known, as well as a variety of ideas related to that notion. In the 1970's, in the Scandinavian countries as well as in Germany, it is in fact possible to locate a richness of experimental work in education, including mathematics education, which could be identified as being truely 'critical'.[3] None of these initiatives, however, identified critical education as its unifying idea. The development of a theoretical framework and the development of examples had an *ad hoc* nature. Nevertheless, I find it reasonable to label this movement as the first wave of critical mathematics education.[4]

In three shorter books[5] I have discussed the notion of critical mathematics education, and in *Kritik, Undervisning og Matematik* I give a more systematic exposition by pointing out three key-terms of critical mathematics education. Not only teachers but also children and students are attributed a *critical competence*, which is considered as a resource to be developed further through their participation in the educational process. The content of subject matter cannot be taken for granted on the basis of established tradition but must be continually subjected to revision. There must be a *critical distance* from the curriculum. The teaching–learning process should be oriented towards the goal of providing students with opportunities to develop their critical competence in the form of qualifications necessary for their participation in further democratisation processes in society. Insofar as both teachers and students are critically oriented to traditional content and subject matter of education, in order to develop their critical competence in focussing on problems outside the educational universe, there should be support for their *critical engagement* in common educational and social endeavours. This sketchy summary of key-terms is naturally too superficial and cannot be taken to provide a basis of critical mathematics education.[6]

By the second wave in critical mathematics education I refer to different trends. For instance, the notion of ethnomathematics, as coined by Ubiratan D'Ambrosio and developed further by among others Paulus Gerdes, Marcelo Borba and Geraldo Pompeu Junior; and the notion of 'criticalmathematics' education, as described by Marilyn Frankenstein, Arthur Powell and John Volmink. The work of Marilyn Frankenstein is an example of the persistent attempts that have been made to give education a form that can help people to relearn mathematics in a way that is not suppressive, but can serve them in interpreting daily life experiences. Frankenstein's work is developed with reference to Freire. The notions of anti-racism, anti-sexism and anti-imperialism have a special position in the second wave, while the political dimension of mathematics education has been expressed differently and often more implicitly in the first wave. The second wave has been very rich and varied and some of the notions have been developed to their limits: for instance the work of Renuka Vithal shows that the concept of 'ethnomathematics' has to be developed in new directions to serve as a progressive force in mathematics education if, for example, we have a post-apartheid situation in South Africa in mind.[7] However, the thematic approach I am going to describe now belongs to the first wave.

2. PLANNING A THEMATIC APPROACH

The two teachers, Marianne Klöcker and Inge Lise Kristoffersen, and I intended the project to run for one and a half months, but actually it took nearly two months (six lessons per week). We were looking for a contextualisation, or thematisation, of parts of elementary mathematics fulfilling conditions such as the following: (1) The topic either has to be well known to the children or it must be possible to describe it in non-mathematical terms. It may be a topic belonging to the daily life situations of the children, but it need not be so if it can be expounded and discussed in natural language. It is important to avoid topics where the meaning of the subject matter can only be explained by developing the whole subject. (2) It must be possible for the children to enter the subject at different levels and to develop the theme appropriately even if their abilities are very different. The theme must not have a specific level of threshold. No 'setting' or grouping according to 'ability' of the children can be accepted as a condition for thematic work. (3) The theme must possess a value of its own. It must not degenerate into being merely an illustrative introduction to a new piece of mathematical theory. (4) Working with the theme must create mathematical concepts, ideas about systematisation or ideas of where and how to use mathematics. And it must develop some mathematical skills. (Clearly, such demands are insufficient for a proper investigation of the 'essence' of critical mathematics education, being only guidelines for the planning process, but this does not exclude the possibility that the actual experimental work may include qualities not contained in its sources of inspiration.)

In the thematic work, we took into account the difference between 'concreteness' in a physical sense and 'concreteness' in a social sense. In elementary mathematics education it is often emphasised that learning has to be situated in what is concrete and, arising from the genetic epistemology developed by Jean Piaget, this has been interpreted as physical concreteness.[8] Mathematical understanding is then conceived as being rooted in our manipulation of physical objects and if mathematics education is to be concrete, the learning environment must offer possibilities for children to manipulate, to operate, and to experiment with objects so that, in the process, their mathematical understanding will grow. However, I reject the idea of concretisation if it is simply understood as making abstract mathematical concepts tangible and visible in a physical sense. I prefer the idea of mathematisation which I understand to be the activity of finding systems and regularities in a chaotic daily life situation.

An important asymmetry between concretisation and mathematisation exists. To concretise means to give abstract mathematical terms a more simple and physical interpretation and by doing so, to make them comprehensible to children. This activity of concretisation is reserved for the planners of the curriculum and as such this activity is isolated from the classroom process. To mathematise means to formulate, to systematise and to make judgements about ways of understanding reality, and hence this activity must take place as an integrated part of the learning process. Both the children and the teacher must be involved in the control of this process. Therefore, in the thematic approach we have to look for 'social concreteness'. In fact, I conceive a majority of the developed examples of concrete materials to be used in mathematical education as being abstract from a social point of view, even if they are concrete in a physical sense.

The teachers and I chose a topic tried before: economic relationships. But we tried to give it a new and more provocative turn. We tried to put the child at the centre by drawing three 'concentric circles', the first having to do with the child itself, the second with the child as part of the family, and the third with the child as part of society. We looked at: (1) pocket money, (2) the child benefit allowance, and (3) money needed for equipment of a youth club. The project "Economic Relationships in the World of a Child" is summarised in twelve units, not identical with twelve lessons, since each unit may include one or more lessons.

Unit 1

The children 'received' 10 DKr a week (about one English Pound) in pocket money. We chose 10 DKr because at that time a great aid programme was being run by a church organisation using the slogan 'A 10 Krone note goes a long way' in asking for money for developing countries. The children had seen lots of posters with the slogan and advertisements appeared in all newspapers. Posters had been put up in the two classrooms as well.

The introduction of the first sub-theme was based on a discussion of pocket money: What is the normal amount of money given? How much is reasonable? How do you get your pocket money? Do you have to do something at home to 'earn' it? Is it reasonable to do something to get pocket money? How much work is reasonable? What do you buy with your pocket money? etc. We wanted to push the discussion quite close to the privacy of the children. This was possible because of the teachers' intimate understanding of the children's different backgrounds and because of the parents' sympathy with what was going on in school; they had great confidence in the two teachers. Important, also, was to challenge any implicit set of values saying that more pocket money, acquired in an easy way, is preferable to working for it. Also, it must be noted that exploring the subject of pocket money does not mean intruding into a subject secret to the children. Actually, they talk a lot about pocket money. But if pocket money is a subject suitable for discussion only in the schoolyard and on the playing fields, it may be touched upon in quite an improper way: children may be boasting about how much they get, for instance. Instead we wanted to challenge their implicit opinions.

This discussion was followed by worksheets asking the children to write down what they wanted to buy for 10 DKr, and they were asked to draw a picture of what they wanted to make it easier for comparison. And naturally they had to find out if they in fact would be able to buy what they wanted for 10 DKr. Here we had a small problem in our planning. If the calculation were to be realistic, it was necessary to use decimals which had not yet been explained in the textbook used, but we decided not to take this problem into consideration.

Unit 2

This unit had to do with savings. The children had to choose a more expensive item than one which they could afford to buy. They were asked to find out first, what they wanted, next, the price of the item, and finally, how long it would take to save that amount of money. As homework, the children were asked to find out the real price of what they wanted and they had to elaborate a reasonable plan for how they would save for it.

Unit 3

If it turned out that more or less similar items had different prices in different shops, a discussion about quality, fashion, supply and demand would be useful. In this unit they had to compare the results of their savings and to return to questioning the amount of pocket money. This was related to questions about salary and income. What is a reasonable salary? Is it right that some people earn more money than other people? What are the reasons for differences in salary? We wanted to get close to these political questions as an introduction to the work about child benefits.[9]

Unit 4

The child benefit allowance in Denmark has undergone different changes. At the time the project took place, the child benefit depended on a variety of factors, but in the project we made the simplification that each child received the same amount of money in child benefit, 450 DKr every third month. This unit was introduced with questions like: Do you know if your parents get the child benefit? How much money do you think they get? Does every parent get the child benefit? What is the reason for having a child benefit? Is it possible for you to discuss with your parents what the money is used for?

The money was paid to the children in a 'real' way (the school had 'counterfeit' money) to let the children feel the money 'between their fingers' and so to get the impression that it was 'a lot of money'. Then a restrictive condition was made: The money had to be used for buying clothes. Would it be possible to buy what is needed for a child for the coming springtime and summer for the sum of 450 DKr? The first activity was to figure out what sort of clothes everybody wanted, and, of course, price had to be taken into account. We wanted the activity to be as realistic as possible. As homework, the children had to find out what clothes they already had, what they needed, what they would probably get from elder sisters or brothers and what they wanted.

Unit 5

The children had to draw up a list of what they needed and wanted and then to guess the price of the different items. Afterwards they had to find out the real prices. A great variety of catalogues from different warehouses had been collected with the help of the children and their parents. And the most difficult thing, we expected, would be to stick to the maximum 450 DKr.

Unit 6

A trip to the shopping centre could be necessary.

Unit 7

The final shopping list was elaborated, and it was illustrated by drawings or cut-outs from the catalogues to make a poster exhibition of what had to be bought.

Unit 8

A little statistical investigation based on the exhibition was carried out. This investigation presupposed that some different categories of clothes had been agreed upon, such as: shorts, pullovers, Levis, etc. The result of the statistical investigation, put into diagram, became a starting point for a discussion of fashion and quality.

Unit 9

In this unit the economic perspective was broadened by looking at the child as a member of society. We chose the idea of equipment for a youth club because a new centre was going to be built close to the school. This sub-theme was introduced by questions like: Is anybody able to tell us about what it is like to stay at a youth club? What do you normally do? At what time do you arrive in the morning? Why is it necessary for some children to stay in the youth club? Is it voluntary to go to a club? Who decides? What happens if no youth clubs exist? Why is a new youth club built just around the corner? All these questions touch upon the position of children in society. If both parents are working it could be necessary for the child to go to a youth club after school, and perhaps also an hour or two before school starts in the morning. This situation must be related to the life of families in general. Often both parents have to work. Why is this the case?

Then the work concentrated upon equipment for a youth club: If you were to have the possibility of deciding, how should a new youth club be equipped? Who decides in fact about the equipment? To simplify, we concentrated upon buying toys. By making contact with the relevant authorities in the City Council we found what a realistic amount of money would be. But we had to simplify further by concentrating on what children conceive of as toys. For instance, painting and paper are not seen as toys in the same way as a football is. We imagined that it would be more difficult for the children to imagine how much painting and how much paper it might be necessary to have, but more relevant to discuss the number and quality of footballs. So we calculated the amount of money which could be used for toys in the child's interpretation of the word. It was 8,000 DKr.

This part of the thematic work became real project work. In the first sub-theme concerning pocket money, the children worked on their own. In the next sub-theme, concerning the child benefit, they worked closely with each other, sitting in groups looking in catalogues, making posters etc., but they were normally occupied with their own problems like: Which dress do I really want? But from now on, group work would become essential. Each group must figure out what they wanted to buy for the youth club and to come to general agreement over their suggestion.

Unit 10

A great variety of catalogues had been collected. A difficult problem to solve was how to use the catalogues, especially those addressed to the planners and heads of the centres. They were organised in a complicated way. A main catalogue described all the items which the specific company had to offer: different colours, different sizes, different qualities, different versions of the same sort of item, etc. In the main catalogue a seven digit number was attached to each item, and then the price had to be picked up from the actual price list found in a separate booklet (prices change more often than items).

Would it be possible for the children to handle this complicated set of information? Would chaos result? Would the teaching be hindered by lots of questions about where to find the prices, how to find them? We considered elaborating a more simple list of items and prices but gave up because it seemed to become too extensive a job. We decided to see what would happen.

Unit 11

This became a very lengthy unit in terms of time. The groups had to make out the final list of their priorities. The different members of each group had to agree, and the sum of the suggested expenses must not exceed 8,000 DKr.

Unit 12

The suggestions from the groups were compared and the final version from the class elaborated. It was our intention that a suggestion from the children should be sent to the authorities, but the 'results' of the project work became different.

The mathematical content of the thematic work throughout the units is rather obvious, although not dealt with explicitly in the description of each of the twelve units: Adding numbers (including decimals) up to a certain sum, subtraction, and rough estimation. The work is also about handling huge amounts of data. Some graphical representation is needed, not only in the comparison of what the children wanted to buy but also, for instance, in finding out the results of different ways of saving. From a formal point of view, lots of mathematics was involved.

3. COMMENTS ON THE PROJECT

It was the two teachers and I who took the overall decision that our thematic approach would concern economy and that the three sub-themes should concern pocket money, the child benefit, and equipment for a youth club. Does this mean that the children's possibilities for making their own decisions were eliminated? To some extent the answer must be 'yes'. But the fact that we took the overall decision made it clear to the children that we had some aims and ideas. And that is a new situation compared with teaching by following a textbook. Often a textbook just presents what has to be learned in series of commands made audible, but not understandable, by the teacher. The children's questions of 'why?' are ignored (and, after a period, forgotten by the children themselves) or they are seen as an interruption to the teaching. The introduction of our theme should have made obvious to the children that it made sense to ask 'why?'. This did not mean that the children immediately accepted the ideas presented, but a new teaching–learning situation was brought into existence because the 'why-discussion' could take place inside the horizon of the children. Normally, when a teacher, following a textbook, makes serious attempts to give reasons for what the children have to do, he or she gets into problems. A consequence of the structuralist approach in mathematics education is that the reasons behind what is going on come forward only later on in the educational process, after the developed concepts obtain a complexity making them applicable and which gives them a meaning outside their pure logical structure.

I can conceive of all sorts of public stratifications of children to be destructive for their learning processes. This may not necessarily be the case for all children, but it is for children not able to establish that instrumentalist attitude which fits into 'school logic'. Children have to be taught and to be engaged in the same thematic work in a meaningful way. They have to act and to interact on the same subject. Some children cannot be treated as 'foolish' by offering them simple and, therefore, insulting exercises. These demands mean heavy conditions for a thematic approach. It must be possible to be engaged in the thematic work at different levels, but still in a way which makes sense in relation to the overall aim for the thematic work. When this happens it is realised immediately by the children, and in the present project real engagement was shown, even from the children who normally were seen as 'less able'. Public stratification is a source of personal fatalism although not the only source. Such stratification makes it necessary for lots of children to try to 'explain' to themselves why they are not able to learn this and that. We attempted to eliminate all sorts of public stratification in the thematic work. This does not mean that children do not make comparisons, but to compare suggestions for a youth club is quite different to comparing abilities in calculating numbers.

One phenomenon which surprised us was the general increase of the children's abilities. In connection with the sub-theme 'pocket money' they had to add decimals to find out how much they could buy for 10 DKr and to calculate savings. As mentioned, they had not yet met this in their textbooks. Usually, the implication would be that this had to be explained first and exercises done by children before applications could be worked out. We wanted to see what was going to happen. And what did happen? Nothing at all! The children just found out what they wanted to buy, added the prices and found the sum. This contrasts with usual mathematical lessons using textbooks. Each new step has to be explained, each new page. The children expect the teacher to explain. They expect not to know what to do. They have learned this is the way it is to be. Mathematics has to be learned in the 'proper' way. But this 'knowledge' was forgotten during the thematic work, and the children's abilities increased remarkably.

We observed the same phenomenon in connection with the youth club example. As mentioned, we used some complicated catalogues describing facilities for kindergartens and youth clubs. We feared a total breakdown into a confusion of questions but, again, nothing happened. The children found out. I see two reasons for this. First, when a child already knows about a subject it is easier to improve knowledge and to use intuition adequately. When, for instance, they found that a football costs 2,500 DKr, they knew something had gone wrong. They were able to correct their thinking, and they also wanted to do so. Second, a child's ability increases much when he or she tries to do something which he or she really wants to do. There is an enormous difference between 'has to find out' and 'wants to find out'. When the orientation is decided by the child, an epistemic 'energy' is released. Because it was important to work out the cost of the equipment chosen for the youth club, calculation abilities increased.[10] An important condition for the engagement of the children was the actual relevance of the subject matter discussed. The introductions to the three sub-themes were close to their private interests and were treated seriously. Naturally, seriousness need not always be interpreted like this; seriousness could also mean 'serious fun'. But seriousness was a condition for children to be involved in the different tasks. It was also important that the teachers expressed indignation concerning economic inequality, for instance about inequality of income. It was not our intention to make the subject matter appear neutral but to make visible some of its sensitive political aspects. This was an attempt to give the children an opportunity to situate themselves in a process of mathematisation.

Normally it is idealistic to expect an educational process to turn into (collective) action outside the classroom, but we tried to develop this possibility by challenging the children's attitudes towards spending pocket money and their attitudes towards fashion clothes, and by making it possible for the children to present suggestions to relevant authorities concerning what to buy for a youth club.

4. THE DIARY AND THE 'RESULTS'

One of the two teachers kept a diary, which gives an impression of the life in the classroom. I quote from the diary:

20/3

We talked about the posters and I 'informed' the children that they were going to get 10 DKr per week in pocket money. They worked with interest and confidence. Their drawings were good. Also the less able children were interested. They felt they could cope with it . . .

21/3

Their suggestions for what to buy with their savings became overwhelming: Stereo, a Ferrari, . . . Jens-Erik wanted to buy a Ferrari. 'But, Jens-Erik, you won't be able to drive it yourself.' – 'Before I have saved enough money, I will be old enough to get a driver's licence.' Lars saved 10 DKr each week. In one year it became 499 DKr. 'How did you get this amount of money?' – 'I am so sorry, but I bought sweets for 21 DKr.' . . . We had to use some catalogues to check the prices. Nobody went to the Shopping Centre or called the shops, except Kaj.

22/3

We talked about what they wanted to buy. Still, some were unrealistic. I asked them to work out new suggestions. Kaj couldn't get the information from the radio shop. They wouldn't give it to him. We talked about differences in quality, and about reasons why new things always are invented. They were interested in hearing the different proposals. They were quiet and followed attentively.

27/3

We started to talk about the child benefit and about buying clothes. Jens had asked several times: 'When are we going to start?' Jens-Erik knew about the child benefit, and also that it was intended to help children. Jens was sure that the reason for getting different amounts of money was difference in salaries.

They started to write down what they wanted. It is going to be too abstract: 'I do not need anything.' I asked them to fill out their work sheets as homework. So, now I am curious about tomorrow.

28/3

The most frustrated from yesterday continued: 'I do not need anything.' ... When they got the catalogues (from the different shopping centres) things began to work. They enjoyed looking in the catalogues and finding the prices. Some made much use of the table of contents in the catalogue from Daells Varehus.

29/3

They balanced costs so as not to use more than 450 DKr, which did not cause problems. They started working on the posters. They concentrated hard. They were cutting enthusiastically.

30/3

Some had to join a rehearsal. The rest continued to cut and paste.

3/4

We made a statistical investigation. They listed, and I wrote on the blackboard the types of items of interest ... They managed somehow to draw column diagrams. We had some problem of space. Twenty children cannot move around on four square meters (in front of the posters) without friction. Later, they compared and talked quietly together. – I am becoming a bit tired of the organisation. This forever running about talking and my trying to find the right answers. Perhaps they are a bit too young, too busy and noisy.

24/4

Youth club. We used only ten minutes on the introductory discussion. They were a bit giddy and silly (returning back from a short holiday). Filling out the worksheets was good. They worked in a more concentrated way and most enthusiastically. They were very objective and discussed fervently whether they had to buy twelve or twenty skipping ropes . . .

To agree before they write things down was not simple. One group found it impossible. They did not want to take that trouble. But later they had to realise the advantages of collaborating . . .

25/4

The work with the catalogues proceeds fine. Some children work seriously, others less seriously. A great diversity exists. Control is impossible.

26/4

Working with catalogues.

27/4

Working with catalogues.

4/5

Working out the balance has begun. The children are working very differently, so the introduction for each lesson is troublesome. They have troubles in their group work. We used the calculator – it was very motivating.

The diary did not continue any longer. As mentioned, we intended the work of the children to end up with more specific suggestions concerning equipment for the new youth club. The class, which the diary was about, did not finish by writing a letter to the authorities containing the children's proposals. The other class, however, wrote letters to the principal of an institution, integrating kindergarten as well as a youth club, near to the school. We had got a lot of information and useful help from the principal in our planning of the project. The letters became very short, and not all the children wrote a letter. The writing of letters was not about work as intended, but turned out to become personal:

Dear Vibeke,
It has been nice work and great fun to look in the catalogues. It was difficult to agree. Somebody found it hard to use a calculator. But we found it fine. Now we have to say good bye.

Best wishes from Christian and Henrik

Dear Vibeke,
I think we have had a fine time, because of working as we did. We have learned a lot. I find it most fun to use the calculator. Also, I find it fun to look in the catalogues to find out the prices.

Greetings from Karina

Dear Vibeke,
I find it interesting to have the opportunity to use my head. Also I find it interesting to look at the catalogues.

Greetings from Stephen

Dear Vibeke,
In our group we find the theme interesting because we can do different things. We have worked on buying clothes and about a youth club. I find working on the youth club to be the best because it is most fun.

Rikke

Dear Vibeke,
I hope you are fine. At the moment we are working on something which does not have to cost more that 8,000 DKr, and on what has to be bought for a new youth club. We have nice weather. At the moment sunshine, just fine for taking a walk to the lake.

Best wishes from Marianne

Dear Vibeke,
I hope you are OK. . . . We have nice weather. Just right for taking a trip to the lake. Finish.

Best wishes Lotte

Once more we have learned that it is impossible to predict the results of experimental teaching.

5. EXEMPLARITY

It is obvious that the project "Economic Relationships in the World of a Child" includes many aspects, but which of these are essential in critical mathematics education? Is it important that the project relates to daily life experiences of the children? Or that the discussion relates to more fundamental economic questions? What about the activity of the children? And what about the fourth claim of the thematic approach: working with the theme must create mathematical concepts, ideas about systematisation or ideas of where and how to use mathematics? Is this claim fulfilled? And to what extent does the fulfilling of such a claim becomes essential to critical education? It is not sufficient that we may have a feeling that the project had been 'good', we must try to understand what this might mean. It is also obvious that this is not a straightforward empirical question: What to look for? Besides, how do the previously developed general concepts like democratic competence, mathemacy and critical citizenship relate to the project? Such general concepts are not immediately operational in relation to particular educational situations.

I shall restrict myself to explaining a bit more about one notion which may help us in doing some theoretical bridge building. That is the notion of *exemplarity* which in fact has its origin outside the critical trend in education. This construct is essential for understanding some of the sources of inspiration for the notions of 'project work' and 'thematic approach' as elements of a critical education. 'Exemplarity' also becomes a link between general educational considerations and educational practice. The idea of exemplarity has been a part of the 'metaphysics' which made us see "Economic Relationships in the World of a Child" also as an entry to a more general understanding of economic features of society.

A conference on secondary and upper secondary education was held in Tübingen in Germany in 1951. One of the main questions at the conference was how to realise aims of a general (or liberal) education in a situation in which the curriculum seems to be ever-expanding, where new subjects and new information become incessantly added. If all important 'information' should be handed over to the students, every exploration of particular questions and subjects seems impossible. The danger is obvious but how can we prevent making education the superfluous teaching of a multitude of routines in which the students are not really involved? At that time the concern for a general education was at the top of the agenda of German education and the concept of *Allgemeinbildung* became central.

One possibility for overcoming the problem of the expanding number of subjects is to identify some basic structures and with those structures as guiding principles, to elaborate a curriculum. According to this structuralism, as advocated among others by Jerome S. Bruner, the basic features of a subject can be grasped without splitting the logical connections of the subject into a mass of information. This well known approach has guided general educational reforms during the 1960's and, together with Bourbakism, also that of mathematics education. However, a different point of view was formulated at the conference in Tübingen by means of which it was suggested the problem of the expanding curriculum could be solved. The expression 'exemplarity' was not actually used in the resolution from the conference but the main idea was made clear: fundamental cultural values can be experienced and understood on the basis of the concentrated study of an example. A main part of the German education at that time was influenced by 'Geisteswissenschaftliche Pädagogik', and education was seen as an introduction to the values of the humanities and other important human cultural products. Naturally, mathematics can be seen as a most important human construction and therefore mathematics education has to introduce the students to the cultural values integrated in that subject. [11]

One of the participants of the conference was Martin Wagenschein, and after the conference he became very active in developing the idea of exemplarity in mathematics education and in the natural sciences. In the inter-war period Wagenschein had been a teacher at an experimental school, which did not organise teaching by the usual timetable but in 'periods'. Such a form of organisation may also have supported the conception of exemplary education. Wagenschein did not develop a single well elaborated exposition of his ideas but they are outlined in different papers, so Wagenschein's conception of exemplarity has to be brought together from different parts of his work.

I shall summarise by means of three epistemic assumptions (not enumerated explicitly by Wagenschein). The assumptions concern the epistemic object, the relationship between the epistemic object and subject, and finally the epistemic subject. First the epistemic object: a specific phenomenon can reflect a totality. This idea condenses a holism: a total complexity can be present in a single aspect of that complexity. Wagenschein exemplifies this in different ways: a fundamental feature of human history can be present in a single historical event; a particular natural phenomenon can comprise a whole set of natural phenomena; and a single mathematical proof can encompass a rich system of mathematical proofs. Wagenschein emphasises that the individual phenomenon is not a step towards a totality but is a mirror of that totality.[12]

The second thesis concerns the relationship between the object of knowledge and the knowing subject. It says that it is possible to come to understand a whole complexity by concentrating on a particular aspect. It is not only possible that the individual phenomenon can be a mirror of a totality, it is also possible for the subject to grasp the totality by concentrating on the singular. A profound understanding of human history can be obtained by studying a specific event. The general can be comprehended via the particular.[13] Wagenschein emphasises that the idea of exemplarity is the opposite of specialising; he does not look for simplifications but for the complexity in the particular. Wagenschein illustrates this thesis several times when, for instance, he tries to show how Euclidean geometry can be developed on the basis of a single proof. He tries to show that by concentrating on the theorem of Pythagoras students can obtain a global understanding of geometry.[14] Wagenschein does not find that students working with Pythagoras' theorem become isolated and narrow-minded. Instead a well-chosen question can provide the entrance to a whole subject and getting deeper into a certain subject means, still according to Wagenschein, making contact with other subjects and general principles of science. To Wagenschein, interdisciplinarity is the result of an absorption which starts from a particular but 'opening' question.

The third thesis about the epistemic subject states that it is possible for a subject in his or her 'totality' to be caught, 'shaken' and absorbed by a specific question, and to be completely engaged in the process of 'coming to know'. Therefore education, based on the principle of exemplarity, must arouse the curiosity of the students and proceed from opening questions and problems towards the interdisciplinary grasp of 'values of humanity'. By these formulations Wagenschein reveals that he also embraces Romanticism, which has been a dominant part of German culture.

The three theses unite the principle of exemplarity. A specific example can become a bridge between the epistemic subject and the object. This educational conjecture is not based on empirical evidence in any traditional way, although it is based on Wagenschein's interpretations of his own experiences. The advantage of the notion of 'exemplarity' is that it extends between the concrete situations of educational practice and philosophical interpretations of 'reality', 'knowledge' and 'person'.

Oskar Negt has further developed the idea of exemplarity but in a different context. Negt is interested in vocational education, and in *Soziologische Phantasie und exemplarisches Lernen* he tries to reevaluate some of the ideas of exemplarity and to relate them to the concept of 'sociological imagination' as developed by the anthropologist C. Wright Mills. More fundamental, however, is the inspiration Negt has had from Critical Theory and from Adorno. Adorno never tried to express any specific educational consequences of his critique of education. He restricted his critical activity to a philosophical interpretation of the term.

Through his negative dialectics, Adorno revealed discrepancies between the conceptual content and the reality of education, but he never tried to articulate any idealised reality. Therefore, he could not set up aims for a critical education – nor did he want to do so. However, Negt gets closer to educational practice.

I shall not try to rephrase Negt's ideas, only to indicate how reformulating the theses of exemplarity may provide inspiration for a critical education. The first thesis maintains that a particular phenomenon can be a mirror of a total complexity. This thesis can have a sociological interpretation meaning that an individual socio-political event (for instance an incident at a place of work) can reflect a political totality. This can be stated in Marxian terms, as done by Negt, saying that the structure of a capitalist economy becomes a basic condition for the life of the individual. The 'totality' need not be defined, as Wagenschein did, as being clear of every sort of cultural conflict and crisis and presented as 'rosy', but can be interpreted as a daily life controversy. Exemplarity can be reformulated and come to refer to critical features of society.

Reformulating the next thesis we find that it is possible to understand a social complexity by concentrating on a particular event. This makes good sense in relation to a vocational education which includes helping the workers to come to know and understand their own political situation. To achieve this, it is not necessary to teach the workers series of basic facts; instead an education with such an aim can start from the particular, meaning that it can begin as a discussion of an actual situation of the workplace. This does not mean limiting the possibilities of coming to know basic features of society. The particular situation reveals general structures and these can be comprehended by concentrating on the particular. This shows why Negt relates 'exemplarity' and 'social imagination' in the title of his book. Here we find epistemic roots for the idea of concentrating on the situation of the learners as a source of what to learn. (It is obvious that Freire has based his approach on a similar idea but without using the terminology of exemplarity.)

The last thesis is more difficult to reformulate, but it can be related to the conviction that the aim of education is not just to transfer information but also to involve the workers in trying to improve their own situation. The education of the workers also becomes a political activity, according to Negt, and that idea was generalised by the first strong movement of progressive education in the late 1960's. The social imagination is oriented towards what actual social reality can be changed into, and education acquires a political perspective. We can use the third thesis as a reminder about taking the subjective dimension of education into account.

Negt's radicalisation of 'exemplarity' can be generalised not just to vocational education and adults, but to education in general. It is not difficult to locate this generalisation in catchwords like 'problem orientation', 'project work' and 'thematisation'. – Why problem orientation? Because a specific problem can become the point of entry to a complexity; a totality can be made comprehensible by an intensive study of a central problem. Project work becomes a possibility when the curriculum of education is not bound by a sequence of logically identified pieces of information. Thematisation becomes yet another way for an educational organisation, not to split the timetable into different subjects, but to organise it in themes.[15] It is possible to look at the ideas connected with exemplarity as a way of making a specification of what could be an educational meaning of *Mündigkeit*. The reconstruction of 'exemplarity' shows that this term from Adorno's vocabulary is related to the concept of 'social imagination'.

Looking back at the guidelines initially set up for identifying the content of a thematic approach, we find the idea of exemplarity expressed in them, although not in an explicit way. When the child in "Economic Relationships in the World of a Child" was located in the concentric circles, we see the influence of exemplary thinking. Pocket money, the child benefit and money needed for equipment for a youth club all have to do with economic structures around a child, and one reason for choosing such 'circles' was that it is important for the child to learn something about his or her position in society. In this way, the aspect of exemplarity, as interpreted by Negt, became present. Not much consideration, however, was paid to the aspect of exemplarity as Wagenschein interpreted the term. The possibility that through a specific example we shall be able to go deeply into mathematical questions was not explored during the project. In fact, the conception of exemplarity as developed by Negt, and to a large extent adopted by critical education, indicates that the original ideas of Wagenschein are not important. Critical education must be concerned with 'social imagery', and a traditional view of mathematical competence does not seem related to this term. This is one of the reasons that the development of critical mathematics education was placed outside the mainstream of critical education.

Only a weak connection between general educational theories and educational practice may be located by the notion of 'exemplarity'. What has been said about the formatting power of mathematics, mathematics as part of technology, democratic competence, and conditions of a critical citizenship, still belongs to the cloudy parts of this investigation. However, the term 'exemplarity' has been the first step in making it possible to find a way from the concept of *Mündigkeit* to the classroom.

CHAPTER 5

"GOLFPARKEN" AND "CONSTRUCTIONS"

Aalborg Friskole is a small private school established in 1981 in an attempt to realise some of the more progressive pedagogical ideas which had been much discussed in Denmark since the beginning of the 1970's, but which also had been difficult to realise in the big comprehensive schools geared to normal education. At Aalborg Friskole, much experience has been gained in organising project work, on both a large and a small scale and at different levels. One subject, however, had never been integrated into the interdisciplinary features to any considerable extent: mathematics. This subject had been thought to be too difficult to integrate in real project work, so in mathematical lessons the teaching followed ordinary textbooks; and while project work and thematic approaches were organised in different subjects and as interdisciplinary enterprises, mathematics remained the reserved onlooker.

One of the purposes behind two projects recently carried out at Aalborg Friskole, "Golfparken" and "Constructions", was to create a broad educational context, rich in possibilities for activities, for cooperative work between children and for the children to decide about their own tasks – and, as something new, within that richness, to develop mathematics. The main aim was not to concretise mathematics but to see in what way mathematics could develop from a broad context requiring the use of mathematics. If mathematics is everywhere in daily life situations, then it need not be necessary to develop any artificial concretisation. Instead the prerequisite must be to create open-ended situations and in these, to let mathematics grow. The approach in planning "Golfparken" and "Constructions" was to use the already broad experience among the staff at the school in organising project work.

After describing "Golfparken" and "Constructions", I shall try to draw attention to notions which might be useful for a further analysis of educational situations in order to bring to light some of those features which might be relevant for critical mathematics education. "Golfparken" and "Constructions" incorporate some of those educational priorities which a broad experience in planning project work in primary and secondary schools had already brought about. In that sense, the two projects illustrate an updated practical experience in progressive education in Denmark, and I see that experience as important for a

80   CHAPTER 5

further analysis of critical mathematics education.

The children involved in the projects were between 10 and 12 years old. The projects integrated about 45 children from three different classes, meaning that the difference in ages between the children was greater than in a normal teaching situation. It was a 'mixed ability' group, i.e. no streaming or setting had taken place. "Golfparken" and "Constructions" ran consecutively for a period of two and a half months in total, with an interruption of a one-week holiday and a two-week period of more traditionally organised teaching. The time used for the projects was twelve hours per week and the rest of the time the children had their normal lessons. The teachers involved in the projects were Jens Jørgen Andersen, Ole Dyhr and Thue Ørberg. Andreas Reinholt from Aalborg Teacher Training College was also involved, and he coordinated the participation of students from the college.

1. OPINIONS ABOUT MATHEMATICS

Before the projects started some of the children were interviewed.[1] It was not the intention of the interviews to try to test any preformulated and specific theses about children's perceptions of mathematics but simply to get some impression of the opinions and emotions which the children had in relation to the subject. The intention was also to give the children an opportunity to express views about mathematics. It could be expected that showing an interest in the children's opinions might help to change their views of their own position in the mathematical classroom from being receivers of information into being participants whose opinions are worth listening to.

The interviews all followed the same general pattern: first the children were given the opportunity to express their general opinions about mathematics, then some questions followed addressing how they saw mathematics in society, starting with a question about mathematics and engineering, and through questions having to do with farming, they were asked about the jobs of their parents and if they knew whether their parents used mathematics. This section of the interviews concluded by asking if the children connected mathematics with anything in their own future. The interviews were open-ended and could take a variety of directions, which they in fact did. The true–false ideology, so widespread in mathematics, saying that every problem when formulated in mathematical terminology has one and only one correct answer, was also addressed.

In general, the children had a strong expectation that somebody, in fact, was using mathematics, but it was obvious that none of the children had ever noticed any adult using mathematics in his or her job or in private. Usually, the children felt sure that an engineer had to do mathematics in designing something like a bridge, but when asked about the farmer's possible use of mathematics they were not sure. Two of the children suggested the idea that certainly a farmer could use mathematics if he or she wanted to do so. The same answer was given when they were asked if they thought that they were going to use mathematics in their future vocational or private life.

Mathematics is not seen as an important tool but as a sort of game which has to be played according to certain standards, and which you can more or less deliberately decide to take up. For the children, the real application of mathematics outside school practice is a mysterious phenomenon – they have never witnessed such an activity. And even if they may have seen mathematical activities, they have not been able to recognise them as such. They do not know what they have to look for, but still they share the opinion that mathematics is an important subject. Because they had never seen mathematics in use, it was difficult for the children to express more specific ideas about that phenomenon. Normally, it is not a part of mathematics education to support the development of a language about mathematics. Meta-level considerations are ignored. However, as the interviews indicate, children have some ideas about mathematics, but they do not get many chances to modify them or to develop them as part of mathematics education.

The interviewer also asked if a mathematical problem always has one and only one correct answer; or if it is possible sometimes to find different answers, all of them correct. The children's answers concerning the latter possibility were definitive: 'No. I've never seen that.' And: 'No, I can't remember such a situation. The true–false ideology is widely accepted, and therefore communication in the classroom cannot be seen by the children as essential in developing ideas and understanding. As long as the true–false ideology dominates, the children do not need to pay special attention to the communicative aspect of the mathematical classroom – primarily, communication becomes meaningful if it becomes difficult for the children to find a solution to a problem. Communication becomes a means of control.

Naturally, these interviews provide only a smattering of empirical evidence concerning children's meta-conception of mathematics, but they can serve as an illustration of two theses concerning such a meta-conception. The first has to do with the social position of mathematics, and the interviews indicate that this position is 'invisible' and incomprehensible to children in general. They find it difficult to express ideas about the position of mathematics and about the actual use of mathematics. They find it difficult to identify mathematics in practical situations and 'at work'. It is important to relate the idea of the invisibility of mathematics to the assumption about the formatting power of mathematics, because if both assumptions are correct, we witness a challenging and critical situation for mathematics education. This conflict has been formulated as the paradox of relevance: on the one hand, mathematics has a pervasive social influence and, on the other hand, students and children are unable to recognise this relevance.[2] The second thesis illustrated by the interviews is that the children hold an absolutistic perspective on mathematics. A mathematical problem is related to one and only one correct answer. That this meta-conception might be misleading has nothing especially to do with the emergence in mathematical philosophy of a quasi-empiricism, and of a scepticistic and fallibilistic trend, which reveal the classical absolutism as problematic. Children's absolutism may concern not only the reliability of 'pure' mathematics but also the application of mathematics ascribing infallibility to real problem solving where mathematics has been used as a tool. This extended form of absolutism is a real problem in education (even though quasi-empiricism might be wrong).

The fact that children have great difficulties in identifying any use of mathematics outside the classroom, also has an implication for the potential for motivating children. Sometimes general motivation is based on the possibility that a child can see himself or herself in a specific and attractive situation in later life, but that sort of motivation is nearly impossible to develop in mathematics education. The difficulties of identifying mathematics in use also have a consequence for the application of the idea of exemplarity, as developed in terms of Negt. To some extent it is presupposed that it should be possible to grasp the general sociological perspective of a specific example. This means that the original outline of exemplarity in education presupposes a sort of transparency. It should be possible, according to Negt, to relate the specific task in the educational process to general social problems. But this assumption about transparency is dubious when mathematics is concerned.

2. "GOLFPARKEN"

Next to Aalborg Friskole is a small natural park called Golfparken. This became the focus of the first thematic work. The project was divided into different units, not of the same length; each unit could vary from about one hour to nearly one week. The project was planned primarily to involve creative art and design in its first stage, then through biology it could move onto mathematics, which was seen as the 'emerging subject'. If the following description looks well-organised, it was not quite the case – naturally lots of confusion occured and interruptions took place. The following units give only an outline of the main direction of the project.

Unit 1

A brief talk about Golfparken introduced the project. Golfparken is not a very interesting piece of land. What could be done to transform it into a fascinating area, into a landscape which the children would like to have in the neighbourhood of the school? Some of the ideas that came up were: Make it into a wood, and why not with palms? Or a grove of fruit trees – apples, pears and plums? It could become a lawn with lots of possibilities for football, or it could contain trees with lianas. It could also be turned into a swamp – water is always beautiful in children's eyes. Or a cave. Or it could be a beach, and why not make it into a real leisure centre, which must naturally have a water chute.

An important idea of this introduction was to release the children's imagination from all sorts of practical restrictions. The question was not what to suggest as being realistic (in light of, for instance, the authorities' lack of money and different rules and regulations) but to create a free space for fantasy to flourish. The enthusiasm was great but, at the same time, it was difficult for the children to get hold of the 'direction' of the whole project.

Unit 2

During this unit the children had to work in pairs to build up a fantasy landscape. Each group received a large piece of cardboard, and on that they made a montage of scenery using cuttings from magazines, shiny paper, pieces of cloth, etc., and a lot of glue, of course. For some of the children it was difficult to get hold of the task: What is meant by fantasy landscape? This unit lasted two days and when it was finished, the various suggestions were brought together into a single picture and put on the wall of the classroom. The different displays seemed to unite into a real adventure landscape.

Unit 3

Animals were going to be placed in the 'landscape', so the children got a new piece of cardboard containing two ovals drawn as the basic pattern for the body of an animal, and using this as a start they created their own animals. The ovals were produced to give a start and immediately they raised questions: Should they be coloured? Is this the body and that the head? Should they become covered by the fur from that old frock? (The children had lots of discarded rags and junk at their disposal.) The animals were cut out from the cardboard and placed in the adventure landscape.

At first some of the children lost their concentration: – What is all this going to be? Then the stories about the animals began to take form: – One morning when an Omaplixila-dragon flew out to find some swamp and moss it bumped into an Okyko sitting in a tree ... The children made suggestions about how to continue such stories. It was also obvious, simply by looking at the patchwork landscape on the wall of the classroom, that the animals were going to interact.

Unit 4

A questionnaire was distributed to the children. It contained items concerning the animals produced. How big? Their height? Their weight? How strong are they? What do they eat? How much? Are they carnivorous or herbivorous? Do they live in the trees? Or in the swamp? Can they swim? Do they have wings? How fast can they fly? To several of the children the first step in finding answers was difficult. Where to look up an answer? Then things took a turn: – If Hanne's and Peter's animal is able to fly with a speed of *that*, our animal must be able to obtain a speed of at least ... ! And besides it eats twice as much! All the information was put into the questionnaire and photo-copies were made showing the animal and its data together. That made it easier to look up information when needed.[3]

Unit 5

The children worked in groups to write stories about their animals. The original idea was that each group should write a story together, but some children preferred to do their own version. When the stories were finished the children had the opportunity to make a 'professional' edition by using a word processor, and they enjoyed seeing their stories in print. The stories were collected in a small book, and each child got a copy of the fairy tales.

Unit 6

An environmental adviser from the local authorities was invited to the school. He talked about his job and together with the children he visited Golfparken. Here he explained the meaning of showing consideration for the environment, and the children came to know about the complexity of nature and of the richness of the 'microscopic' life in the area. Some of the children listened most attentively, while others were engaged in other minor distractions. There is a small lake in Golfparken into which stones and sticks can easily be thrown. Among some of the boys there was the usual pulling and pushing – just to see if the nearest person would take the next (dangerous) step closer to the water.

Unit 7

Later the children were divided into two large groups, one to look at the life in the small lake in the Golfparken, the other to investigate the countryside. Several instruments had been borrowed by the school for the work: magnifying glasses, pipettes, landing nets, a pH-measurer, an oxygen-measurer, etc. The children had to catch as many of the different sorts of insects as possible. Every time they caught one, they had to fill out a work sheet with different details about the insect in order to determine its name. A big diagram drawn on waterproof material made it possible for the children to determine the names of most of the insects. The children had to follow the tree-diagram and to answer different questions: How many legs does the insect have? Does it have wings? And finally at the end of the branch they could find the name of the insect.

Unit 8

The children were also given a new questionnaire with the same structure as the one used when the information about the animals from the adventure park was collected. The children had to fill out the questionnaire: first the name of the insect, then several other bits of information as, for instance, if the insect was found in the lake, whether it was found on the surface or near the bottom. The questionnaire demanded two kinds of information. First, what it was possible to see (number of legs, etc.) and next, what it was possible to find out by reading about the animals. A collection of books was available in the classroom. A part of the questionnaire could be completed as field work, although sometimes the wind took hold of the papers.

Some of the children returned to the school with wet socks. Every small lake seems to demand the test: How far is it possible to walk out, before the water starts running over the tops of rubber boots?

Unit 9

This unit consisted of a trip to the zoo in Aalborg with the special purpose of coming to know about the life of mice and about their behaviour. At that time a special exhibition about mice was held at the zoo, and that little animal was also one of the inhabitants of Golfparken.

Unit 10

This lengthy unit had to do with tabulating numbers of all the collected material and information. Computers were available and the children recorded the different information. They also had to make diagrams of the lake and indicate the different living areas of the animals and insects. Many diagrams and statistics could be produced, and the children had seen a rationale for this work by being involved in the whole task of collecting the data.

An important experience for the children was to see the relevance of having the information put on a computer. It became possible to ask questions like: How many insects did we find living on the surface of the lake? How many did we find living at the bottom? How many were living in between? A clarification of such questions could bring about a new one: Do those numbers tell us anything about the quality of the water in the lake?

Unit 11

In parallel with the work of organising data, the children produced a little story about an animal living in Golfparken. The stories had the following beginning: 'I am a small water flea living ...'. The children had to identify themselves with one of the small animals and to write about life from that perspective.

Unit 12

The project finished by trying to return to the original questions, now put into a realistic context. What could be done to Golfparken in order to transform it into an attractive area?

3. "CONSTRUCTIONS"

Mathematics did not in any way dominate "Golfparken" even though, as mentioned, the intention was to make it possible for mathematics to grow out of the richness and openness of the educational organisation. In "Constructions", mathematics was planned so as to play a more crucial role. Mathematics can be used in different ways; it can be used in parallel with other means of description. In "Golfparken" a situation was created which called for several types of description, in "Constructions", however, mathematics was thought of as a tool for design. It makes a difference in using mathematics whether you have collected sets of data and want to use mathematics in an attempt to illustrate some relationships or some correlations, or whether you are trying to make some sort of construction. The second project tried to look at mathematics from this constructive point of view.[4] "Constructions" did not take the form of a thematic approach but a variety of different workshops were set up. The groups of children then had to rotate from one workshop to the next.

The day before the children started on "Constructions" a parents' evening was arranged. That meant that from the very beginning the parents got a proper idea of what the project was about and what the workshops involved. The workshops were set up in the great hall of the school and the parents had an opportunity to try them out.

Workshop 1

The task of this workshop was to build and construct using Lego. It was also possible to use the advanced system of Lego-Technic, including possibilities for making connections with a computer and other steering facilities. First, the children had to accomplish a specific construction following a step-by-step plan. Later, they could make experiments and make their own constructions; however, it took more time to finish the step-by-step construction than expected. Often the boys just started building and playing with the Lego, while the girls more carefully tried to finish the task of making the planned construction.

Much mathematics is involved in such construction practice: counting, proportionality, measuring, etc. Because a plan is in two dimensions and the construction is three dimensional, much 'if–then' reasoning has to take place: – If that is 'double' and that it covered by the red 'triple', then we have to use two of that sort first in order to . . . , etc.

Workshop 2

Computers were available in this workshop, and the children worked in pairs. They were asked to experiment with a Danish variant of LOGO, and the first task was to control the manoeuvre of the mouse on the screen. Afterwards they had to give orders to the mouse by writing small programmes. Later their work became more advanced. The children concentrated on different things: some carefully calculated the angle at which the mouse on the screen had to be turned and the distance it had to move in order to draw a specific pattern, while others were happy to use the information and ideas they could glean from their friends.

In this workshop it was also possible to observe a difference between the attitudes of boys and girls. While the boys tended to rush towards the more complex and interesting tasks, not caring too much about the more routine work, the girls were often careful in first trying to master the elementary skills, which meant that when they got to the more challenging tasks, they had a headstart. It was also interesting to notice the ambivalence of the children when some other groups used one of their ideas. The children became a little flattered if another group picked up an idea, but at the same time ideas were conceived of as being personal property: 'It is our idea, don't use it!'

An interesting and most important observation was that the explicit calculation of angles and lengths were normally carried out when something had gone wrong in the more intuitive way of handling the problem. Mistakes can be seen as a strong motivation for using explicit mathematics. Explicit calculations are the consequence when intuitive ideas become insufficient, and mathematics becomes the language in which to discuss the nature of mistakes.

Workshop 3

This involved building with sticks 15 cm long and with a diameter of 1 mm. The first task was: Build whatever you want, and you are welcome to cut the sticks into smaller pieces. Next: Build a diamond made up by triangles. And finally: Build a cube. The children automatically realised the necessity of supporting braces when the construction did not contain triangles.

After the construction of the cube they were given the opportunity to build whatever they wanted, and by chance the idea to build a plane by covering a skeleton of sticks with thin paper spread amongst the groups. A competition to make the best plane started. The planes were tested, not only in the hall but also outside the school, and that created a bit more space in the overcrowded hall. One of the children who had had little success in the normal lessons happened to construct one of the best planes and this gave him a lot of prestige which was his for the rest of the week.

Workshop 4

The main task of this workshop was to construct a kite by using a workshop drawing of a model. A main object of this workshop was to learn to interpret such a drawing. When finished, the children tested the kites. This also helped to reduce the overall 'density' in the hall.

Workshop 5

This had to do with measuring, and the children were assigned some specific tasks: measure yourself, your height, wrist, waist, ankle, etc. After that they had to measure different objects from the school – the door, the window, the table, etc. – and finally, different distances in Aalborg by using a map. One child took a special interest in this workshop and was helping the other groups in measuring.

Workshop 6

In this workshop the children had to cut out different geometric forms from polystyrene. The first task was to cut out a box and count the edges, faces and the vertices. Next: Cut off one vertex of the box and count the number of edges, faces and vertices of the new brick. And this was to be continued. The following task was to cut out a cube and divide it into two 'triangular' objects, and that exercise was also followed by questions. One task was to make a box the same shape as a Toblerone bar and to find out about the angles – and several more tasks were set up.
    The cutting of the packing foam was a bit difficult to handle. A heated wire was used for cutting but too often the wire broke and the main job for the teacher became to substitute the broken wires while the children lost concentration and interest.

Workshop 7

This also involved constructions with sticks as in Workshop 3, but now the sticks were bigger, the length 50 cm and the diameter 1 cm. This workshop also involved different tasks, the final one of which was to make a bridge with a span of at least 1 m. The children got some ideas from looking at a picture of a railway bridge and picked up the basic principle that the sides of the bridge had to be made up of triangles. The stability of the constructions were tested by hanging a bucket in the middle of the bridge; then water was poured into the bucket until the construction seemed to reach its limit. The amount of water was then a measure of the strength of the bridge. Not surprisingly a competition arose to make the strongest bridge. And of course some of the bridges suddenly 'did not win'.
    The original intention was to have a competition to make the longest bridge but that idea was changed into a united project of building a

tower. As part of the project "Constructions" the children had visited the (well-known) tower in Aalborg which is of steel construction. Not all the bridges had collapsed during the bucket-test, and some could be repaired, so a sufficient number could be collected to build a tower. This task became the final spectacular undertaking of "Constructions". The tower was made up of the different one-meter bridges and reached a considerable height. An elevator, made as part of the Lego-project, was installed, and it could go right to the top of the tower which, in fact, had an appearance similar to that of the real tower in Aalborg. The tower and the elevator were shown and demonstrated to the rest of the school during a school assembly.

4. COMMENTS ON THE PROJECTS

Long discussions were held between the teachers concerning both the details and the more general aims of the projects. One of the teachers expressed the main problem as the fact that the mathematical content was not developed very much, and with regard to "Golfparken" he added: "When I discussed the outcome of the project with the children in my class they first said that they did not see any mathematics in the project. Then we discussed the different situations when they had to calculate and to use the computer. Well, perhaps some mathematics, they thought, but mostly biology." Other teachers agreed but one added that it depends on how you look at mathematics: "Perhaps it is not so important whether the children feel they are working with mathematical problems or not, if they in fact are doing so." Nevertheless, it was generally agreed that the project did not really get into mathematics, and one teacher summarised this by saying: "We did not finish the work with diagrams and statistics, I find this a little annoying. More mathematics should have been worked out."

It is obvious that the teachers were not satisfied with the mathematical content of "Golfparken". The idea of mathematics as an emerging subject had not been realised. One reason could have been that the situations created did not contain possibilities sufficient for the application of mathematics. However, many sorts of descriptive uses of mathematics were not only possible but also most appropriate in "Golfparken". Another possibility mentioned was that the children in fact used some mathematics but were just simply unaware of it – and perhaps not even the teachers had been sufficiently aware of this fact. But is it true that if the children had used mathematics implicitly, then it does not matter whether they realise it or not? And if it does matter, what is the next step to be taken? Should "Golfparken" in some way have been prolonged? The same question could be raised about "Constructions".

I shall outline some notions to which we shall return later on. *Setting up a scene* for an educational process refers to the effort of establishing a situation into which the educational process can be embedded to provide the individual activities of the children with a kind of meaning. The scene should make it possible for the children to see motives for the different activities and to verbalise what sort of competence could be developed.[5] In "Golfparken" the scene was set first by the introduction of a fantasy landscape followed by concentration on the actual Golfparken, and in "Constructions" by the establishment of the different workshops (although in this case the scene-setting seems different). I do not use the term 'setting up a scene' for isolated ideas and suggestions which may catch the attention of the children for single lessons, but I have in mind a general outlook which provides a perspective during a longer educational process. It is important that the scene gives the children possibilities for grasping an idea of what has to be done and for what purpose the specific tasks could be. The expression 'setting a scene' indicates that something artificial takes place, and only in rare cases is it possible to avoid this in an educational situation. A scene should not be mixed up with reality.

The setting up a scene took place in the project "Economic Relationships in the World of a Child", described in Chapter 4, by specifying some everyday situations and problems to which the learning activities were related. Actually, scenes were set up three times by the three sub-projects: pocket money, the child benefit, and equipment for a youth club. Before the overall project was introduced the children were asked to bring catalogues from different department stores, and they started asking when they had to begin looking for what they were to buy. They were curious about what it all could be about. Then the first actual setting up of a scene took place in the discussion of pocket money; the next in the discussion of the child benefit; and the last in the discussion of the use of youth clubs. The settings had contents which are serious and important not only from the actual perspective of the child, but also from a sociological perspective. Children's use of money is an important subject, and the same is true about the child benefit and the child's position in society made apparent through the need for youth clubs. In "Economic Relationships in the World of a Child" the children made suggestions for what to do with child benefits but did not actually buy anything. In this sense the social embedding was a game. Nevertheless, a scene-setting provides connections between the different tasks which the children have to do and it provides 'meanings'.

A scene can be set up in different ways and looking at the described examples we find three different ways of doing it.[6] First there is the realistic approach, exemplified by "Economic Relationships in the World of a Child". The material used was realistic, the prices real, and the problem actual. But realism is not the only strategy, and in "Golfparken' the first scene was set up in an atmosphere of fantasy and fairy tale. What was to be done with the landscape outside the school? The children were invited to make their suggestions without the limitations of 'reality'. Everything was possible, but the children were not dropped into an empty space because a scene was set up. The scene-setting made it possible for (most of) the children to identify a purpose for their tasks in relation to a general aim. Naturally, it would have been possible to use a realistic approach in the first phase of the project "Golfparken" as well. A third way of setting up a scene is illustrated by the variety of workshops – creating a 'working-together atmosphere' in "Constructions". In this case the scene-setting involves the children by means of 'engagement' and 'activity'.

A much discussed question in relation to project organisation and the problem orientation of an educational process is: Who is going to decide what is the initial problem – the teacher or the children? In fact all scene-settings described in "Economic Relationships in the World of a Child", "Golfparken" and "Constructions" were made by the teachers. But how essential is 'self-organisation of the educational process'? It need not be the case that the children actually accept being engaged in, for instance, making suggestions for buying equipment for a youth club – and in fact that sub-project finished a bit earlier than the original plan and scene-setting had suggested.

A fundamental idea I shall connect with scene-setting is that of providing *meaning* for an educational process. Some answers to the children's question: 'Why are we going to do that?' may be indicated by the scene-setting. For instance, the different calculations to be performed in order to make a suggestion for buying equipment acquire a meaning in relation to the scene. The children are able to locate a meaning if the scene is set up in a way they find comprehensible. But it is not claimed that the meaning the children locate is *the* meaning or *their* meaning. Even if the children find a meaning comprehensible, it need not be accepted. It does not make sense to say that a scene may provide *the* meaning, only that it provides *some* meaning.[7] To try to provide a meaning is not the same as demanding that the children should accept that meaning. Meaning is an offer, not a claim. Not all scene-settings are successful in establishing a meaning: In the first phase of "Golfparken" the children expressed confusion, they did not see any point in investigating a fantasy landscape.

It is tempting to compare this with traditional mathematics teaching in which the textbook often provides the principal structure. The individual lesson only acquires meaning by being a step in finishing a well specified curriculum task. It could be maintained that in all educational situations a scene will be set up, and that a textbook normally defines a scene, but I shall not use the concept in that liberal way. I want to restrict the use of 'setting up a scene' to examples such as those referred to, implying that scene-setting presupposes that a specific perspective is introduced to which the activities of the children can be related. The normal meta-language of traditional education is governed by a 'logic of command'. Setting up a scene acknowledges the importance of children being able to see a meaning for their individual educational tasks – and this is not just a question of motivation. A scene-setting is a way of breaking the 'logic of command' expressed by the sequence of exercises along which every child is guided towards an invisible goal.

Meaning in education is related to the availability of a *language about* what is done during an educational process. Educational meaning is related to a meta-language and to the possibilities of discussing alternative directions of an educational process. Therefore, it is important for the children to get hold not only of the subject matter to be learned, but also of a meta-language of what has to be learned and for what purpose. It is important not only for the children to grasp a meaning but also to have the possibility of negotiating a meaning about the content of their education.[8] This means that I see the notions of 'setting a scene', 'meaning in education' and 'meta-language about educational content' as being closely related. The importance of developing such a language about mathematics is also indicated by the interviews which initiated "Golfparken" and "Constructions". The children had opinions about mathematics but no adequate means of expressing them.

A scene-setting is an attempt to provide a language, not identical with the language of the subject matter itself, to identify what is done and what has to be done. We just have to imagine the difference between children discussing what to do with respect to certain exercises in a normal textbook and children involved in calculations in one of the sub-projects of "Economic Relationships in the World of a Child". But the demands of a meta-language also have a different perspective. Learners are acting persons, and an action cannot take place without reason. To take a well-grounded decision presupposes clarification and negotiation, it presupposes a language in which it is possible to describe and discuss the content and aims of the different alternatives.[9]

The idea of negotiating a meaning and acquiring a (meta)language can be clarified further by the term *multi-faceted semantics* (or multi-faceted pragmatics).[10] Let us take a look at a simple concept such as 'addition of decimals'. This concept has many contexts. From the perspective of a child the concept may be related to the addition of integers, maybe it is also related to calculations using a pocket calculator, it can be related to activities performed in previous lessons, etc. It is also a part of the daily life of the child: addition of money for instance. But the daily life concept of addition of decimals need not be governed by the logic which we expect a mathematical concept to obey. For the child the algorithm of addition may depend on the units in question. It is different to add up amounts of money, at least when the amounts are at the pocket money level, from adding weights for instance. Money can be visualised in different ways than weight, and a visualisation may create a certain algorithm. The concept of addition is located in a network of semantic relationships. Leaving the perspective of a child aside, we find that a mathematical concept still has a variety of semantical relationships, some settled in different sorts of work practice and some in pure mathematics. The semantical network is alive, it is a multi-armed creature which could be envied by an octopus. Scene-setting is a way of acknowledging this fact and an attempt to include such semantical richness in the educational process.

As indicated by the remarks from the teachers, mathematics seems to have disappeared in "Golfparken", although this project had incorporated plenty of possibilities for bringing mathematics in use. The children had opportunities to use a variety of mathematical competencies but did not recognise that they had performed any mathematics, and this fact leads us towards the importance of a *mathematical archaeology*.[11] This concept becomes interesting because a scene-setting also may mean a changed focus. Not only does it provide possibilities for educational focuses but also for a systematic distortion of the focuses. Mathematics in use is not a transparent phenomenon and therefore exemplarity is not a straightforward task for mathematics education.

Mathematics may be integrated to such a degree that it disappears for both the children and the teachers. Then it becomes important that time is spent on getting hold of the imbedded competence. Mathematics has to be recognised and named, that is the task of a mathematical archaeology. But the projects "Golfparken" and "Constructions" finished without such an archaeology – too early, so to speak. It is important that a project which contains mathematics as an implicit element does not end at the moment when projects normally do, i.e. when the most visible parts of the projects are produced and the results are exhibited. Some time spent on mathematical archaeology could be useful. And why is this important? If the children in fact were concerned with mathematics, and if they in fact had learned mathematics implicitly, why then be bother with a mathematical archaeology? Why is it important from the perspective of critical education? My first answer will be something like: If it is important to draw attention to the fact that mathematics is part of our daily life, then it also becomes important to provide children with a means for identifying and expressing this phenomenon.

An aim of a mathematical archaeology is to make explicit the actual use of mathematics hidden in social structures and routines. It is the process of digging mathematics out and drawing attention to how mathematics moves from being an explicit guide to becoming a grey eminence underlying, for instance, social and economic management. A mathematical archaeology is an attempt to explicate thinking abstractions which confront us as realised abstraction – it is a response to the integration of mathematics into our second nature. This is a global interpretation, but mathematical archaeology can also be given an educational interpretation.

As part of the work in "Constructions" the children constructed a bridge, and they realised that when they used the long sticks to make cubes no stability was established. A much better solution to the problem of stability was to use triangles. It could have been of importance to go further into the question about the construction of the bridge. In elementary education a mathematical archaeology could mean an illustration of the fact that engineers, involved in technological constructions, have to apply mathematics if they are to avoid similar troubles to those of the children. The educational activity could have involved a visit to a workplace for engineers, looking at drawings, looking at figures with calculations, etc. Naturally, the idea is not that the children should understand the mathematics in use, but that mathematics has been in use. The actual example also provides opportunities for talking about geometry in a more concrete way: Given three lengths it is possible to construct one and only one triangle (if one of the lengths is not too long compared to the others). From this experience it would make sense to discuss further geometrical concepts to illustrate that these concepts could be applied when talking about the constructions they had just made, and that the concepts could be used in making improvements of those constructions.[12]

To be able to see what mathematics in fact is doing means that mathematics must first be identified. A contextualisation may be more or less artificial and more or less complex. If it is so complex that mathematics seems to disappear, then it makes sense to follow it with a mathematical archaeology.

A mathematical archaeology is an educational activity which involves being aware that some activities carried out in the classroom – for instance the way of systematising information about different insects found in Golfparken – are in fact mathematics. Further, uncovering the mathematical roots of an activity is related to the idea of the formatting power of mathematics. If bringing about a discussion of the power of mathematics is to make sense, then we have to be able to identify examples of mathematics in use. Therefore, a mathematical archaeology can make sense both when directed towards activities in a classroom and towards social phenomena. The problem is the same: applications of mathematics are difficult to observe and therefore to express an opinion on. If they stay invisible and beneath the technological surface of society, they get out of control. When the children fail to realise that mathematics is in action, they do not have any chance to question their own opinions about it. When they do not realise that they are using mathematics, their image of the subject as belonging only to a textbook is not challenged.

CHAPTER 6

REFLECTIVE KNOWING

The term 'mathemacy' was introduced as being comparable with that of 'literacy', but the competence of mathemacy has not yet been explored. The task now is to find an interpretation of mathemacy which can identify connections between the content of experimental educational practice and lofty concepts such as 'critical citizenship', by means of which I have tried to characterise an overall position of critical education. Adorno never tried to apply his exposition of education to the classroom. He did not try to identify aspects of a critical education within an educational practice. His critique was conceptual and only that. Some bridging is necessary and the concept of 'exemplarity' has already been mentioned, but this concept has mainly had the function of producing ideas for practice by identifying a thematic approach and project work as possibilities; 'exemplarity' does not explain what such organising principles have to do with a critical education.

We have seen examples of different educational activities, but what features of these activities are essential to a critical mathematics education: The integration of different subjects using a thematic approach? The political significance of the questions related to subject matter as in "Economic Relationships in the World of a Child"? Or the possibility for children to be involved in a variety of activities as in "Golfparken" and "Constructions"? Is it a problem that the children, in "Golfparken" for instance, did not seem to be aware of their mathematical learning? The general concepts outlined so far seem not to be able to isolate features of educational practice essential to a critical perspective. The lack of connection between 'a philosophy of' and 'an educational practice' has simply not yet been overcome.

It is essential to bridge the gap between philosophy and practice because the notion of critical mathematics education cannot be derived from one of these alone. I find it impossible to construct an idea of critical mathematics education by using an Adorno type approach only, nor could the conception of critical practice be formed as a compendium of a variety of well-meant activities. To ask what features of a practice may be essential for a critical education calls both the concept of critical mathematics education and that of educational practice into question. Therefore, I shall now try to locate an epistemic concept by means of which we may be able to find potential for educational practice in our theoretical constructs, as well as parts of the general perspective in practical situations. Using the notion of reflection, I shall try to put more content into mathemacy.

1. REFLECTIVE KNOWING: A FIRST DELINEATION

We cannot predict the consequences of the way technological development influences social life. We live on top of unknown risk structures – that is the reality behind the Vico paradox. In today's highly technological societies, formal methods and technologies play a major role in the creation and constituting of those structures which beset actual social development. Together with technology, which contains frozen mathematics, we are situated in conflict zones between constructive and destructive potentials. From an epistemological point of view this situation calls for an important distinction.

Let us, as an example, look at the problem of motoring. Too many (private) cars cause pollution. This form of transport carries some serious risks (of an ecological nature, for instance) the consequences of which we are going to face in the not-too-distant future. The way to confront these emerging problems is not to develop the driver's driving skills, i.e. their ability to manoeuvre a car in traffic, nor is it to give the driver more information about mechanics – how the car is actually constructed, how the brakes work, how they can be repaired, and so on. Naturally, it is useful both to be able to repair a car and to drive it in a better way, but this is not a satisfactory answer to the problem of motoring presented. To face this problem and to react to it in an adequate way, we have to develop a better understanding of motoring, seen as the complex phenomenon of organising transport and traffic in general. What are the economical and ecological consequences of motoring? What social and political actions are needed and which seem to be possible? We have to learn *about* the activity of motoring as a whole to address such questions. Obviously, to learn about motoring is not itself a solution to the problems caused by engaging in it, but it is the epistemological step to be taken to get hold of the problem itself. Knowledge at a meta-level must be developed if our actions are not to degenerate into measures of desperation. To stick to the improvement of driving skills would be to adopt an ostrich-like policy.

Let us call the knowledge necessary for developing and using a technology *technological knowledge*, and, as an example, we may include both the knowledge necessary for driving a car and the know-how necessary for repairing and constructing it. Driving skills are not of the same type of knowledge as knowledge about (the activity of) motoring. The latter is an example of meta-knowledge, and for the moment I call it *reflective knowledge*. However, the concept of knowledge has to be broadened. Technological knowledge includes not only explicit and verbalised knowledge, in the usual interpretation of the word, but also a variety of competencies to perform different acts – motoring ability is not only explicit knowledge, some of it may be organised as tacit competencies. This complex of competencies I call simply technological knowledge, and that is the object of reflections which also may include both verbalised and non-verbalised competencies.

The fundamental thesis relating technological and reflective knowledge is that technological knowledge itself is insufficient for predicting and analysing the results and consequences of its own production; reflections building upon different competencies are needed. The competence in constructing a car is not adequate for the evaluation of the social consequences of car production. Improved road ability does not lead to a better understanding of motoring. Technological knowledge is born 'shortsighted' as it were. Reflective knowledge has a broader basis. It must be based on a wider horizon of interpretations and pre-understandings.[1] It has to grasp the situation in which technological knowledge is at work, but no simple step leads from technological constructions to reflections. Technological and reflective knowledge constitute two different types of knowledge, but not two independent types. It may be important to master some technological insight to support reflections. To be able to understand and to discuss social implications of the pollution caused by motoring we have to know about basic principles and conditions related to the construction of cars, but it is not necessary to master all aspects of this construction. If this were the case, democracy in a highly technological society would become impossible: only experts would be able to control experts. The result would be an 'expertocracy'.

Reflective knowledge cannot be analysed into bits of technological knowledge. Even if we collect every available piece of technological information, we shall not be able to build up reflections from these parts alone. Reflections cannot be put together by a rearrangement of technological information. Technological knowledge does not contain within it a self-critique, nor a specification of the alternative trends in technological development. So reflections do not have their epistemological basis in technology, but in the sociological and ethical aspects of the way we handle technology. While technological knowledge aims at solving technological problems, the object of reflection is the complexity of the implications of a suggested technological solution to these problems. Technological language refers to its own domain, while the object of the language of reflective knowledge includes technological solutions to these problems. Reflective knowledge subsumes technological knowledge and language, but refers to norms and values as well.

The importance of reflective knowledge arises from the existence of conflicts between the constructive and destructive impacts of technology. Technology may well be a reaction to some critical aspects of society but in the attempt to solve these, new risk structures are created. We are not solving problems and dissolving risk structures – they are transformed instead. We cannot uphold an optimistic interpretation any longer (technological crises may emerge), and it is impossible to stick to a determinism (critical action is possible). Therefore, we have to develop a fund of reflections of technology and its consequences. It is the conflicts between the constructive and destructive impacts of technology, i.e. the critical potentials of technology, which makes critique necessary.

By means of technological knowledge we may be able to change technology and we may also be able to reflect upon a way of solving a technological problem. But is it possible to change a technology by means of reflective knowledge? In my view, the answer (hopefully) is 'yes', due to the critical ingredients of supposed technological solutions. This means that I see the Vico paradox as an expression of a permanent challenge and not as something impossibile to overcome (in particular situations). The Vico paradox is caused by the impossibility of making an epistemological reduction of reflective knowledge to that of technological knowledge. Had it been possible to build up reflections from bits of technological competence, humanity might be able to cope with its own creations. To justify an optimism in spite of that, depends on the possibility of making some sense of 'reflection'.

In what follows, I shall leave the general perspective of technology to look at those parts of technology in which mathematics plays an essential role. Therefore, it may be necessary to make a further distinction between technological and *mathematical knowledge*. The formal component of mathematical knowledge is integrated into the different technologies and especially into information technology. This implies that we end up by differentiating between three types of knowledge: (1) Mathematical knowledge, which refers to the competencies we normally describe as mathematical skills. (These include competencies in reproducing mathematical thoughts, theorems and proofs, as well as in performing algorithms for calculations. The advanced competence of inventing and discovering new mathematics is also included.) (2) Technological knowledge, which refers to the ability to apply mathematics and formal methods in pursuing technological aims. (We concentrate our discussions on the applications of formal methods because they characterise highly technological societies.) (3) Reflective knowledge, which has to do with the evaluation and general discussion of what is identified as a technological aim, and the social and ethical consequences of pursuing that aim with selected tools.[2]

Reflective knowledge defined in abstract terms is the competence needed to be able to take a justified stand in a discussion of technological questions. In this sense we may relate reflective knowledge to the general competence needed to be able to react as a critical citizen in today's societies.[3] The possibility for the public to be not only subjects, i.e. geared only to receive outputs from the 'system', but also to provide inputs to the 'system', presupposes reflective understanding. The question remains whether reflections may make a reorientation of a technological development.

My terminology has, however, already run into a problem. The concept of 'knowledge' has several philosophical connotations. In classical epistemology from Plato to Russell, it has been assumed that knowledge must be associated with three basic conditions: To claim that a proposition is known means that the proposition is believed, that it can be supported by sufficient reasons, and that it is true.[4] The conceptual association between 'knowledge' and 'truth' points towards an absolutism, and knowledge becomes associated with authority. Maybe this is why we do not have too much difficulty in talking about 'mathematical knowledge' (even if we accept a fallibilistic interpretation of mathematics). Perhaps we can also use the expression 'technological knowledge' although that body of knowledge cannot be depicted in the same way as mathematical knowledge. It may be more appropriate, however, to talk about technological competence, indicating that the competence used need not necessarily be expressed in an explicit way. We often associate knowledge with explicability while technological competence in some situations takes the form of a tacit competence. It seems even more difficult to talk about reflective knowledge; in this case it seems preposterous to refer to any authorised body of knowledge. The expression 'reflective knowledge' indicates the existence of some degree of explicability and perhaps the existence of some sort of authorised body of knowledge, and this is deceptive. We have to look for this particular competence in terms of dispositions and abilities. Therefore, I prefer to talk about reflections as a process and in what follows, I choose to use formulations like reflections, reflecting and *reflective knowing*

102   CHAPTER 6

instead of 'reflective knowledge'. I shall also use the terms 'mathematical knowing' and 'technological knowing'.

2. REFLECTIONS AND MODELLING

To arrive at a more specific interpretation of reflective knowing we shall look at mathematical modelling.[5] We can conceive modelling as a powerful way by means of which mathematics performs formatting. In a modelling process mathematics not only touches reality, but also squeezes and transforms it. Abstractions become realised. Modelling becomes a technological act and a way of inducing systems into reality. By concentrating on modelling the discussion of reflection becomes not only more specific but also more restricted. Nevertheless, mathematical modelling constitutes a key problem in the evaluation of technologies because the seemingly transparent mathematical language creates a full scale Vico paradox.

Let me make a distinction between two main types of modelling: *pointed modelling* and *extended modelling*. In the case of pointed modelling the problem we face is transformed into a formal language, in terms of which we try to solve the original problem. Extended modelling is different. In this case mathematical terminology is not used to describe a specific problem, but it is used to provide a generic foundation for a technological process. Mathematics becomes part of a conceptual framework by means of which we interpret and rearrange reality. As an example, we could think of the distinction between 'value of exchange' and 'value of use', which is fundamental to the capitalistic organisation of economic life and which is realised through the system of double bookkeeping, making it important to establish general calculation of values in terms of an abstract money system.[6] In this case the mathematical calculation model provides a means of transforming and handling a complex situation. In the extended type of modelling, mathematics is absorbed into parts of our basic conceptual system for handling social affairs. Mathematics becomes a transcendental condition for individual phenomena – and also for classes of pointed modelling. The distinction between pointed and extended modelling means that we can also split the concept of formatting and talk about local and global formattings. Here, I shall concentrate on local formattings and try to specify a process leading from thinking abstractions to realised abstractions.

The initial activity in a (pointed) modelling process is to locate a *problem area* and an area of interest. It would be an oversimplification to conceive of a problem as belonging to reality. It makes more sense to interpret a problem as a defect or a lack in a conceptual construction. However, it makes even more sense to identify a problem as an inconsistency between a mental/theoretical construction and a real situation. In other words: the concept 'problem' does not refer to just one of the two categories, language or reality, but designates a relationship between them.[7] To use an economic example: the problem area may have to do with the economic policy of a government: How is it possible to analyse and, to some extent, predict the most likely outcomes of suggestions for an economic policy? Is it possible to indicate some general implications, for instance concerning the number of unemployed, by using techniques of economic forecasting? An attempt to answer such questions is advanced by the development of the 'Simulation Model of the Economic Council' (SMEC) used by Danish economists when advising the government and politicians on economic policy and its possible consequences.[8]

A mathematical model must be based on a specific interpretation of reality. Other possibilities do not exist. We cannot come into contact with a 'reality' without structuring it. This statement has been stressed in the philosophy of science as a reaction to the doctrine of logical positivism, where objectivity and neutrality are seen as aims which are possible, though sometimes difficult, to obtain. N. R. Hanson has explained why observations in general, including scientific observations, will become pre-structured in accordance with our conceptual patterns; it is not possible or even important to try to avoid these patterns.[9] Instead we may use investigations of our pre-structured world as a basis for restructuring activities, perhaps leading to an improved understanding. A model can never be a model of reality. We have to identify elements of reality which are to be conceived as being the important ones; we also have to decide which relationships among these elements are essential. In this way we create a *system*; and this is not a part of reality. The system is conceptual and created by means of certain interpretations of reality, i.e. by means of a certain theoretical framework for looking at reality, and by having certain knowledge-constituting interests in mind. The preconditions for the process of system construction can be implicit or explicit. They can be anticipated or accidental, but they are inevitably present. A conceptual systematisation of reality is a prerequisite for mathematical modelling. To specify micro- or macro-economic structures presupposes decision-making about what the main elements of an economy are and what the basic linkage between the selected elements is. System development cannot be grounded in a simple positive act of observing. This is the reason that I prefer the term 'system development' and not, for instance, 'system identification'.[10]

An attempt can be made to model the developed conceptual system, and in this way a *mathematical model* becomes a specification and a semi-description of a conceptual system created by an interpretation of features of reality. By means of the mathematical model, the verbal and non-formal descriptions of relationships between different parameters may be expressed in terms of functional relationships, and characteristics of the relationships may be expressed as properties of the selected mathematical functions. The mathematical terminology comes to substitute the terminology in which the system was originally outlined. In SMEC the Cobb–Douglas Function of Production plays a crucial role. It determines the Gross National Product as a function of two variables, namely capital investment and the labour force. This specification is a consequence of encompassing the development of the system by neo-classical economic theory. The model is developed with this function as its kernel. The next step is to specify additional mathematical conditions which the function has to fulfil, for instance concerning its differentiability and the properties of its partial derivatives. Such specifications are expressed in the form of the equation of a function, and by solving this, more is learned about the Function of Production. Then, with this function as the point of reference, the model is broken down into a great number of equations, making the model useful for performing economic forecasts.

If a mathematical model is to be numerically manageable, it may be necessary to transform it into *algorithms*. A main question now is how to define the numerical methods required to handle the necessary calculations. Often it is a complex manoeuvre to transform mathematical descriptions into algorithmic procedures. (It should be noted that a mathematical model need not be put into numerical form to be useful. A model can disclose correlations and important features of an interpretation of an economic structure, even when it is not possible to proceed to a numerical stage.) In order actually to use a mathematical model, the formal results have to be *interpreted and realised*, and based on these interpretations, actions may be prescribed and carried out. In the case of SMEC, results are handed over to the politicians and come to influence the political and economic debate, and finally, the decisions that are made.

To summarise, we have identified the following activities as being included in the process of pointed mathematical modelling: *problem identification, system development, mathematisation, algorithmatisation,* and *interpretation* including a *realisation*. These activities have been mentioned in sequential order. It is obvious that several loops are involved in a modelling process, but more important is the fact that no sequential order between these activities need exist at all. It may be the case that a certain model is already available and has been applied, but then a new and different interpretation confronts the model. This makes it necessary to return to the process of system development in order to cross-examine the original preconditions, and then to realise that the problem area could be located in a different way. The mentioned sub-processes of mathematical modelling cannot be put into any sequential order. I have just sketched a mixture of integrated activities.

The described activities of modelling can be seen as a possible specification for a route leading from thinking abstractions into realised abstractions (where we locate interpretations, including a realisation, as the 'final' step in the modelling process).[11] Thinking abstractions are characterised as thought models, while realised abstractions get a different status: they belong to reality itself. Previously, I have tried to specify the route in terms of a transition from one type of formalisation to another – from a formalisation of language to a formalisation of routines. By realising abstractions fundamental features of the 'abstraction society' are established, and this act can be discussed in terms of the elements of mathematical model building: a basic element in realising abstractions is the materialisation of the mathematical specifications of a conceptually developed system. Mathematical modelling comes to exercise a formatting power.[12]

Pointed modelling and extended modelling are mentioned as extremes. Pointed modelling concentrates on well specified problems embedded in a local and 'narrow' system, while an extended use of mathematics is found when a whole area or part of social life becomes the object for a modelling. As indicated, we could think of our monetary system as a jumble of mathematical models and economic ideas. It constitutes a totality of frozen mathematics. If we design a model to make a prognosis for some investments, this modelling will be embedded in a totally mathematised structure which exemplifies an extended model, while the particular model for making the prognosis is pointed. The activities described within a modelling process best fit the pointed modelling. As I have drafted the modelling process, it has become an activity performed by particular groups of people, while the extended modelling is better interpreted as a cultural product; it seems to have a multifarious social origin.

A main problem related to the mathematical modelling process has to do with the concealment of pre-understandings. This is the phenomenon of disguising the complexity of the construction of the conceptual system which constitutes the very foundation of the model itself. Therefore, the identification of pre-understandings becomes a major task for reflection.

It is often forgotten and eliminated during the modelling process that a mathematical model is not merely a model of reality but represents a specific interpretation, based on a more or less elaborated theoretical framework and some interests. Instead, a strange metaphysics develops, ascribing objectivity to the model itself. The language of mathematics makes the very process of system construction invisible, and by doing so it becomes difficult to identify the nature of system development that has preceded the model in question. According to the Cobb–Douglas function, national production is determined by the labour force and capital investment, and a whole metaphysics of economics is built into economic models via this central function reflecting, for instance, the assumption that ecological costs lie outside the framework of the question when economic patterns are described.

Sometimes the process of system development can rely on well-established theories, this being the case if the modelling deals with particular physical phenomena already described in classical mechanics. In this case, the pre-interpretation will be of quite a different nature than if the modelling relates to economy. The aim of a reflection is not to try to eliminate pre-interpretation, but to identify the nature of the understandings which have preceded the mathematical modelling. These pre-understandings are normally obliterated by the mathematical language, making different models look identical if mathematically they appear identical. The mathematical modelling process takes place against a background of pre-understandings. It takes place against a background of interests (political, economic, etc.), depending on the nature and scope of the modelling process. Furthermore, it takes place in a situation characterised by a certain distribution of power as well as a certain distribution of more or less well established theories and prejudices. Structures of interests, powers and theories (or prejudices) make up the background for a modelling process, and the *first task of reflective knowing* can be described by the principle: Try to make explicit the preconditions of a modelling process which become hidden when mathematical language gives it a neutral cover.

According to Russell, Wittgenstein (in the *Tractatus*) and Carnap, the act of formalising a natural language is an important epistemological step.[13] The richness of natural language reveals only confusion and ambiguity, which emerge because the grammar of natural language is inexact. Natural language is therefore insufficient for the formulation of scientific ideas and theories, instead it is a main source of misconception. It becomes necessary to create an improved language which can become the language of science, a project which has been pursued by logical positivism, and especially by Carnap, who saw a formal language as the foundation of the unity of science.[14] It was assumed that a formal language could depict reality because the logical structure of such a language would 'fit' the logical structure of reality. In the *Tractatus*, Wittgenstein expressed a similar view by underlining that the limits of language coincide with the limits of the world, and here language is interpreted as formal language. Natural language descriptions are of inferior quality to formal ones, so a 'translation' of natural language formulation has to take place to separate out the metaphysics built into natural language.

In the article 'The Elimination of Metaphysics through Logical Analysis of Language', Carnap explains in great detail the weakness of the grammar of natural language. It allows the formulation of sentences which, from a grammatical point of view, are correct, but which do not contain any significant information, i.e. sentences which do not state empirical facts or hypotheses. These must be conceived of as pseudo-statements. Carnap therefore suggests that the language of science should be so developed that it accepts as grammatically correct only those sentences which have a proper scientific content. The basis for this language should be the language developed in formal logic. Using this language as a reference, Carnap gives the example of how it is possible to reveal that different propositions from Martin Heidegger's *Was ist Metaphysik* (concerning the Nothing and Anxiety which are key-terms in the ontology of existentialism) cannot be translated into unequivocal logical form. They break several principles of formal grammar. Such propositions look like assertions but, according to Carnap, they are simply nonsense. In this way formal language saves us from a metaphysical quagmire. By purifying our language we come to purify our thoughts, and the limits of our formal language become the limits of our (scientific) world.

In opposition to this view, the 'natural language philosophy' has emphasised that a formal language only gives a very rough and imprecise picture of reality. The richness of natural language is essential because it creates a descriptive potential which can never be achieved by a formal description. In this philosophical tradition, a formalisation is seen as a (primitive) language game built up by a few words (the logical connectives) which may have a 'civil life' in natural language, as it were, but which after having been recruited to do their 'military service' in the ranks of a formal language must behave in a drilled and rigid way. This position states that natural language contains an infinity of possibilities for ascribing meanings and nuances, which is necessary to serious communication. A formal language can only become a dead and very imprecise plaster cast of a part of natural language. This position is maintained by the 'natural language philosophy' as expounded amongst others by Wittgenstein (in the *Philosophical Investigations*), Gilbert Ryle and J. L. Austin. These two opposing positions provide two different ways of looking at the formalisation of language: a formalisation may be a productive epistemological act, or it may be a bombastic simplification.

Connected with this discussion, we find another aspect of the tasks of language. One presumption is developed in the *Tractatus*: a language provides a 'picture of reality'. The picture may be precise or not, but the thesis by which Wittgenstein accompanied the Picture Theory is that a formal language is able to give the best picture of reality possible. In the *Investigations* Wittgenstein refuses to accept this assumption, himself becoming a strong critic of his earlier work and developing a much richer interpretation of the 'jobs of language'. Instead of a theory of language as a picture, he sees language as a collection of 'games', and by so doing he introduces a theory of language as action. The question is no longer whether language provides an adequate picture but it becomes a question of what language game we are playing; and the notion of language games develops into a theory of speech acts.[15]

We often talk about mathematical models according to the Picture Theory: The model describes an object to a certain degree of approximation. However, this way of interpreting models easily repeats the 'fallacy of *Tractatus*', i.e. the idea of language as a picture of reality. It is possible to try to 'overcome' this fallacy, not by changing the model, but by changing reality, by allowing objects described in mathematical language to assume an existence. The mathematical model is then used in a new way. Instead of being a description of something, it becomes the layout of a design. This 'elimination' of the fallacy of the *Tractatus* consists exactly in such a transmission of thinking abstractions into realised abstraction. The problem is that this 'solution' of the fallacy is dealt with in an implicit manner. We still talk about mathematical models as describing something actual, while they in fact prescribe something potential. This phenomenon of prescription adds a deeper meaning to the thesis of linguistic relativism.

I shall take up the idea that we are 'doing' something through the use of our language, and from this perspective I shall take a look at the modelling process. Seeing language as action provides an opening to the normative aspects of the technological applications of mathematics. I shall locate four different language games involved in a modelling process to provide a better understanding of the 'irrationals' in realising abstraction. Specific problems are connected with transitions between these different language games.

First we have the language in which we talk about 'reality' and try to locate a problem area. I do not say that this is a homogeneous language, it may be better described as a conglomerate of different language games, but still this cluster could be given a name, and I shall call it *natural language*. Using this, we try to handle reality. Natural language is based on commonsense interpretations of reality, and these interpretations are not always consistent or well-founded. They may comprise misunderstandings, confusions, parochialism and prejudices.

We can invent a specific terminology and use different theory-loaded terms to highlight aspects of reality, but these aspects do not exist as simple parts of reality. The highlighting provides some parts with a different status and they become part of a system. The language of system development is different from natural language to the extent that it includes technical terms cultivated as part of a theoretical discourse. A transition from a natural language to a *systemic language* takes place as an essential part of system development. This language incorporates terms based on the selected theoretical framework used in an interpretation of reality. If we again take economy as an example, terms from the systemic language may be 'the ideal consumer', 'elasticity of prices', 'government-expenditure multiplier', 'marginal cost', etc. To use such terms makes it possible to 'see' new aspects of economic life, however these aspects are not 'raw facts' but elaborated configurations.[16]

A new step in the modelling process is taken when a mathematisation takes the form of an incomplete transition from a systemic language into a *mathematical language*. Instead of informal descriptions of relationships between certain parameters, specific mathematical functions may be introduced. 'Marginal cost', for instance, becomes defined in terms of the derivative of a function. Depending on how we look at the epistemic potential of a natural language and of a formal language, we may describe the route from natural language to mathematical language as a semantically constructive or destructive trip.

To make the model numerically manageable, algorithms have to be invented, and this step also involves a transition from one language into another. It could be the case that we have to leave the terminology of mathematical analysis and apply the language of numerical analysis. Let us call this an *algorithmic language*. Finally, let me emphasise that it is only in my description that the transitions between the languages games referred to take place in a sequential order. The previous remark, that the sub-processes of mathematical modelling cannot find a linear order, also means that no natural order exists between the transitions from one language game to another.[17]

A modelling process involves linguistic transitions all characterised by being incomplete translations – this is why I prefer the term transition. System development takes the form of a transition of a natural language into a systemic language, while mathematisation becomes a transition from a systemic language into mathematical language, etc. The four mentioned languages games – the natural, the systemic, the mathematical and the algorithmic do not belong to the same linguistic 'type'. They are different with respect to syntax, semantics and pragmatics. The syntax of mathematics is essentially unambiguous while the systemic and the natural language are not. We have to add 'essentially', because the language of mathematics does not accomplish the standards of formalised languages established by the formalist tradition. According to formalism, the grammar of a formal language has to be unambiguous, but mathematics at work must loosen its formal tie. However, the programming languages which take over the algorithmic jobs have unambiguous grammars. Similarly, it is possible to point out differences in the semantics of the different types of languages. Natural language has a semantics which makes it possible to discuss normative questions, while this is impossible in a mathematical language. The transition leading from natural to a mathematical language, therefore, means cutting away semantical possibilities for formulating normative controversies and uncertainties. The systemic language stands in between; it may keep a door open to normative discourse. The pragmatics, understood as the context of the discourse, are also different for the different types of languages. We only have to think of the differences between the social groups for whom the possibility exists to understand and use the involved language games.

No process of translation will get us from one language into the other. It does not make sense to talk about an activity of translation between languages of different types. Instead, the step from one language into another constitutes an irregular jump – a linguistic metamorphosis. The transitions mean an elimination of possibilities for discussion although something else may be gained. We can associate different possibilities of speech acts with the different language games constituting the modelling process. The act of formalising therefore becomes an activity which opens up new possibilities and closes others. It is a movement between different scopes of possibility for action. The thesis that a technological competence does not encompass the capacity for evaluating the effects of the technological constructions can now be interpreted in terms of the impossibility of translation between the different types of language games involved in the modelling process. The language necessary for identifying consequences is neither the mathematical language nor the algorithmic one. The systemic language in between may allow a discussion of consequences, but this language is loaded with theoretical constructs and may be biased in a way different from the bias of a natural language.

We are now able to specify a *second task of reflective knowing*: Reflective knowing must address problems and uncertainties connected with transitions between the different types of language game involved in the mathematical modelling process. A language game has its associated blind spots, and this means that a transition is accompanied by different potential defects, as for instance the elimination of the possibility of a normative discourse. Reflections have to get hold of the uncertainties of such transitions; not in order to solve that problem – this cannot be solved – but in order to create an awareness of the nature of the transitions.

A speech act is an action, so what do we do when we are modelling? The transitions between language games are parts of a (technological) investigation, which in itself is sensitive to a modelling process. What we have in mind in this context is still a pointed modelling as exemplified by the development of SMEC. We can identify four aspects of the context for the modelling process: (1) problem identification,[18] (2) structure of argumentation (providing a rationale for technological actions), (3) social basis for critique and corrections, and (4) scope of possible (technological) actions. We shall take a look at these aspects of the context subsequently.

A core element of the modelling process is a problem, but mathematics is not a neutral appliance in relation to the investigated problem: during a modelling process we can expect the problem to undergo a metamorphosis.[19] Mathematics influences the task of *problem identification*. A (pointed) modelling process starts because of a (technological) problem. This problem becomes redefined within the framework of a conceptual system, and when we apply mathematics to handle the problem described in the systemic language, we again make reformulations. So, when using a mathematical model we should not expect to solve the initial problem. Modelling is an efficient means of 'transport', it takes us to new places. Or as Lakatos has put it: "After Columbus one should not be surprised if one does not solve the problem one has set out to solve."[20] The problem described in natural language is not carried over into the systemic language or further on into the mathematical language in the same form. Instead it becomes decomposed and reformulated, and changes in the course of its linguistic travel.

The analysis of the problem (or problems!) takes place within a certain *structure of argumentation*. This structure I see as a sort of 'logic' for allowable and acceptable arguments in the problem-solving process. This argumentation can take place in a more or less colloquial form. However, mathematical modelling changes not only the language of problem clarification and solution, but also the argumentative aspect of the solving process. In economic affairs this situation is distinctive. The whole creation of macro-economic models raises new possibilities for the analysis of consequences of economic decisions. The models create different settings for economic management, and, in particular, they create new arguments about how to solve economic problems. The different 'logics' which influence interpretations and realisations can also be illustrated by SMEC. In this case we find the following tendency: If the predictions made by the model in any way seem to conform to the government's general ideas and hopes, the outputs of the model are taken as an argument for the scientific basis and relevance of the economic suggestions. On the other hand, if the model does not fit such political programmes, a critique of the assumptions of the system development is addressed. The presumptions of the model are questioned; if not, the prediction is ignored or explained away. Political opponents may naturally use a reverse strategy. This, I identify as a change in the structure of argumentation. Instead of structure of argumentation, we could also talk about a rationale for technological action. When we are concerned with economic questions, we have rationales for economic policy, perhaps pseudo-rationales because the assumptions in the system construction have been disguised. If we apply a mathematical model in an attempt to handle a technological problem, we at the same time apply a tool that is not neutral in relation to the structure or argumentation. This structure is sensitive to linguistic transactions. It is impossible to state, in general, whether or not such a sensitivity is an advantage. The point is not to block the flow of rationales for action. What is important is to conceptualise this history as part of the reflections on mathematical modelling.

According to Karl Popper, a scientific theory must be connected with a set of 'falsifiers', normally not very well-defined. A falsifier is a proposition, the truth of which contradicts the truth of the theory in question. If a falsifier is found to be true, we have to refute the theory. The set of falsifiers must not be empty, if the theory is to have scientific interest. It should be possible to falsify a scientific theory; it must be open to critique. However, these Popperian formulations can be restated in sociological terms: Every theory must be available to a group of persons for whom it is possible to articulate a critique. That is, it should be possible, not only from a logical, but also from a sociological point of view, to criticise a theory. This means it should be possible for a group of persons to articulate a critique of a suggested technological investigation and of the suggested actions authorised by this investigation. The group of persons for whom it is possible in this way to influence a process of investigations, I call the *basis for critique and corrections*. It consists of the 'set of critics'. When talking about mathematical modelling involved in a technological activity, we may instead talk about the social basis for technological action, or the social basis for management. The thesis is: The basis for critique is sensitive to the nature of language games used in a technological problem-solving process, and the language of mathematics especially influences which groups of persons are given the opportunity to manage and review the type of suggested action. The total process of modelling imposes a severe limitation on the number of persons who are able to follow the journey of modelling from the problem area, into the algorithms (placed in the sinister jungle of the computer) and back again, to open air and natural language. If we are going to investigate mathematical model building, one thing to do is to look at the history of the basis for critique. It is important to identify how the 'amount of power' delegated to different groups of persons accompanying the process of (technological) investigations can change as different media for problem solving are used. This is a task for reflection.

In search for the effects of modelling, we could also take a look at the *scope of possible actions* recommended as a result of the technological investigation. Basically, a process of technological investigations results in an action – in contrast to a scientific investigation which results in some sort of verification or falsification. So, when looking at the scope of possible consequences of action we concentrate upon the genuine technological aspects of the process of investigation. Because a modelling process affects its context, an ethical evaluation is necessary, and the area of possible actions constitutes a main object of critique.

Reflecting on a process of (pointed) mathematical modelling includes a study of the effects which the modelling process has on the main aspects of the process of technological investigation, i.e. the effects on problem identification, on the rationale for technological action, on the basis for critique, and on the space which is opened up for technological management. These aspects make up the context for realising abstractions. In this way we come to a *third task of reflective knowing*: Reflections must address the way a mathematical modelling affects the whole context of problem-solving, seen as a technological enterprise. To condense this idea further: Reflective knowing has to identify the formatting powers of mathematics.[21] The description of the third task of reflection also shows that the overall problem of democracy in a highly technological society is closely connected with the possible effects of the modelling process, as this process can influence the basis for critique and correction.

3. REFLECTIVE KNOWING IN EDUCATIONAL PRACTICE

What do the results of the analysis of 'reflective knowing' mean in an educational context? Naturally, we need not maintain that the only meaning which can be attached to reflective knowing is that of being aware of mathematical formattings, uncertainties connected to transitions between the different language games involved in a modelling process and of preconditions hidden behind a formalism. However, it is important to see whether this interpretation of 'reflective knowing' makes educational sense: Is it possible in an elementary educational context to illustrate some of the described tasks of reflective knowing? (One problem is that we do not necessarily have to do with a concept of reflection sufficiently rich to be a part of the educational aim of developing democratic competence and a critical citizenship.) To specify the nature of mathemacy, it becomes important to take a look at reflective knowing at the micro-level, and we have to move into the classroom in search of the source of such a competence. The basic logical characteristics of reflective knowing need not be changed by taking this step. Reflections may still have to do with evaluation of (proposals for) solutions of technological problems (involving mathematics), but the complexity of the technological problem becomes reduced.

In Chapter 4 we discussed the thematic approach "Economic Relationships in the World of a Child". The theme was divided into three sub-projects, the third having to do with equipment for a youth club. Let me indicate what could be meant by mathematical, technological and reflective knowing in such a context. The children had to balance the budget so as not to exceed the amount of 8,000 DKr. In this context mathematical knowing refers to the children's ability to add numbers in the abstract, and to cope with a problem like: How many times is it possible to add 227 to 6,366 and still get a sum less that 8,000? This problem can be solved by purely mathematical means. A technological problem, in my terminology, is of that type which the children actually were concerned with during the project work, such as: If we buy footballs for the rest of the money, how many do we then get? This question is formulated in natural language and with an aim outside mathematics. The children use a technological competence when they plan their purchases. Stepping back from their proposals and taking a look at the suggested solutions involves a reflection on a technological solution. The questions then become: – Is our proposal for buying equipment satisfactory? Is our proposal better than those of the other groups? How can it be that we have been able to buy much more that the other groups? Could we have made a wrong calculation? Have they bought some very expensive (but important) things? etc. By discussing such questions, reflective knowing has a chance to develop in the classroom. Here we see a reason for scene-setting, which may establish a context making reflections relevant. A scene-setting may create a semantical richness which provides openings for a language of reflection not brought about by a mathematical terminology.

The distinction made between mathematical, technological and reflective knowing is analytical. We are not able actually to identify three different sorts of activities leading to the three different sorts of knowledge. Things are mixed up and not distinguishable. It is possible for the children to learn algorithms for handling mathematical problems and fully concentrate on the mathematical aspects of a problem. The children could ask meaningfully: Have we done the calculation right? A mathematical competence can be pursued even if technological aspects are ignored. This has been done in a lot of education, either because of rigid rituals stressing rules and formality as the essence of mathematics, or because of a well-elaborated, structuralistic interpretation of the nature of mathematics. The children could also ask: Have we done the right calculation? which leads to the technological and reflective dimension. Comparing and evaluating different proposals cannot be done inside the framework of calculation and problem-solving techniques; this indicates that the analytical distinction between different types of knowing at school level is meaningful. The distinction has a basis in educational reality.

It is possible for children to draw upon their commonsense knowledge in pursuing a technological aim and to use non-standard techniques. As noted in the comments on "Economic Relationships in the World of a Child", the children's abilities increased surprisingly when they were dealing with problems embedded in their pre-understanding. Naturally, some basic mathematical skills may be appropriate for solving a technological problem, but it is not obvious to what extent these may be necessary. We could consider the use of calculators, which the children enjoyed so much. An actual outcome of the use of calculators was to separate mathematical competence, as the children had been explicitly taught, from the type of competence needed for solving their problems. When the calculator was present, it was no longer necessary to remember any details of the algorithms for addition, subtraction, multiplication or division, instead the important thing became to understand the nature of the problem, and then to use the calculator: – How many footballs can we buy for the money left? Let us count how many times it is possible to add 227 DKr to what has already been spent without getting more than 8,000 DKr. In general terms: the competence to carry out a technological task is not identical with a (formal) mathematical competence.

Even if children get lots of exercise in performing basic algorithms of calculation, they are not much better off with regard to evaluating proposals for buying equipment. No algorithms exist for evaluating different proposals for solutions of technological problems; more is needed in such reflections. On the other hand, the capacity for making such evaluations is not independent of mathematical and technological capacity. It is necessary to take into account the constraints which are built into the situation and which can be expressed in a mathematical way: – Our proposal is not satisfactory because there are too few footballs. We have to buy some more. Well, that means that we must buy fewer skipping ropes. But do you really think that we could save enough money by buying fewer skipping ropes?

We cannot expect to find any linear time-relationships between the development of mathematical, technological and reflective knowing. It is not the case that children first have to acquire an explicit mathematical competence before they are able to use it in solving technological problems, and that a technological competence has to be developed before they can have any reasoned opinion about a proposed solution. The structuralistic movement in mathematical education has supposed that it was necessary to teach the children mathematical order first before they would be able to apply this order. This assumption has been strongly criticised from the applications oriented approach and from the discussions of 'cognition in practice'. Embedded in our practice we solve technological problems without being able to explain the formal technique being used; we operate in a social context.[22] It is also obvious that children have a broad basis for evaluation and forming opinions. Reflective knowing is not without foundation. Education is also a question of developing and making explicit what is already founded.

Nevertheless, even if reflective knowing is part of the child's world, reflective knowing cannot be developed in all types of educational situations. The situation must be open. This is a necessary condition, if reflections are to come to play an active part in the teaching–learning process (but naturally not a sufficient condition). Reflective knowing cannot be transmitted from a stock of well-established reflective knowledge through a teacher-guided process. Situations have to be created which need reflections and which the children find important to reflect upon. This means that scene-setting acquires relevance as an attempt to communicate the importance of reflective thinking. Standard word problems do not establish conditions for reflection because reflections immediately are reduced to: Have we done the calculation right? Is the result presented in a way that will satisfy the teacher? The problem itself has no importance for the children and the actual content of a solution of the problem therefore becomes irrelevant. The only thing important is the instrumentalist value of having solved the exercise in a satisfactory way. This does not mean that children do not reflect upon what is going on in traditional teaching. Children evaluate what they are doing, but normally these sorts of evaluations are doomed to remain in the corners of the child's mind: – This sort of exercise is boring but simple to solve; you just have to do so and so.

The idea I am trying to provide with meaning (not to prove) is: *If mathemacy has a role to play in critical education, similar to but not identical with the role of literacy, then mathemacy must be seen as being composed of different competencies, a mathematical, a technological and a reflective competence. But especially: Reflective knowing has to be developed to provide mathemacy with an element of empowerment.*[23] It is hardly necessary to state that this formulation raises different questions: If reflecting on real mathematical models used in important social affairs is essential, could we then expect reflective knowing to be a proper educational concept? If we try to simplify the criticism of real modelling, is the result that the teacher comes to determine the students' criticism? If we try to connect reflections with the modelling practice of elementary mathematics education, do reflections then have anything noteworthy to do with reflections of 'real' modelling?

The importance of mathemacy as an integrated competence implies that the guiding principles for mathematics education are not any longer to be found in mathematics but in the social context of mathematics. This means a fundamental change in the focus of mathematics education, and I find this essential in any educational reform which tries to establish critical practice. Mathematics education has, as it were, to be raised to a meta-level, which implies that it becomes impossible to find out what must be included in the curriculum by solely making reference to mathematics itself.[24] Developments in mathematics are not together a sufficient reason for educational reform. It is the role of mathematics in society and the possibility of illustrating the actual formatting power of mathematics which has to be taken into consideration.[25] This epistemic change I see as quite fundamental, but no epistemic underpinning of mathematics education pays special attention to mathemacy as a broad integrated concept. For instance, the constructivist approach as developed from the original Piagetian constructivism concentrates attention on the development of mathematical knowledge, and does not say much about the importance of developing a critical conception of the use of mathematics.

4. SIX ENTRY POINTS TO REFLECTIVE KNOWING

Let me describe some entry points to reflective knowing, still to be interpreted at school level, although these will not guarantee access to the centre of 'democratic competence', nor to the concept of reflective knowing in its general formulation related to the evaluation of technologies of society. A *first* set of questions formulated by students and teachers about their work in the mathematics classroom could be: Have we done the calculation right? Have we followed the algorithm accurately? Are there different ways of controlling the calculation? Such questions all address the mathematical aspects of the problem-solving process, and any attempt to answer these questions immediately moves us back into the area of mathematics. Still they may be seen as rudimentary steps in reflecting upon what has been done. In school all meta-reflections seem to be focused on questions of this type. The dominance of such questions also supports the widespread true–false ideology which embodies much school mathematics.

A *second* entry point is found when we ask questions such as: Have we done the right calculation? Is it possible to chose between different algorithms? Is the algorithm reliable in all circumstances? Is it sound? More generally we could ask: Have we used an appropriate algorithm? Is the algorithm the right basis for pursuing our technological aims?

Meta-reflections need not be confined to the correctness and consistency of the methods used, and we could find a *third* entry point. Reflections could have to do with the reliability of the solution in a specific context. Even if the calculations are done correctly and the accountability of the techniques is established, it need not be the case that the result should be trusted. We could ask questions like: Even if we have calculated in the right way and used an algorithm in a consistent way, do we then find a result which we can actually use? Are the results reliable for the purpose we have in mind? This third group of questions begins the attack on the true–false dichotomy and takes into consideration the context of the use of mathematics. They have to do with means and aims. In this case we are looking at the technological aspect, while in the first two steps, we addressed the mathematical tools. To make it possible for students to raise questions of this type in school, it is important that mathematics is contextualised in a way that they see a value in such investigations. Let us go back to the project "Economic Relationships in the World of a Child". Even if children get the right answer to 'How many more footballs could we buy?' by using the method of subsequently adding 227 DKr to 6,366 DKr and counting how many times this can be done before reaching 8,000 DKr, it need not produce a reasonable answer to the original task which was to make suggestions for buying equipment for the youth club. However, because the children were involved in such a task, it became important for them to take the third step in reflection. This would have been quite irrelevant in a non-contextualised situation; the task then would have been to solve a purely mathematical problem – and that would have been all.

This leads us to a *fourth* entry point to developing reflective knowing. We could raise questions like: Is it appropriate to use a formal technique at all? Do we in fact need mathematics? Is it important to introduce a formal method? Could we find an answer without mathematics? Is the result of a mathematical calculation more or less reliable than an intuitive interpretation of the situation in question? These questions draw attention to the fact that formal techniques and mathematics are not necessary tools for achieving a technological aim. In some cases an intuitive way of handling a specific problem may be preferred. It is an important experience for students that they are sometimes able to find out things without mathematics.[26] If mathematics always have to be at work, it becomes a basis for perpetuating the duality between science and humanities of which mathematics education is one of the instigators. The fourth group of questions attacks that variant of the true–false ideology which says that formal methods must be preferred. Formal methods may take us further in some situations but they do not always give an appropriate answer. By contrasting formal techniques with intuitive ones, it becomes possible to see formalisation as only one possible way of handling a problem, and this experience is important in developing reflective knowing. In the project "Golfparken" the teachers were worried about the lack of mathematics in the main part of the thematic work: Where has the mathematics gone? It seems to have disappeared in the middle of project work. But this could also be interpreted as a positive condition for reflection on the relevance of formal tools.

We find a *fifth* entry point to reflective knowing when we look for broader consequences of the use of specific techniques for solving a problem. By entry points three and four the reflections concentrate on the technological aims of the task, but now we look outside the original aim of our action, and we try to find the general implications of pursuing such a task with formal means. How does the application of an algorithm affect our conception of a part of the world? This is the question addressing the formatting power of mathematics in general terms. But how do we handle this question in an educational context? In taking our fifth step, we must not forget that we are still working within a classroom. However, before returning to this question, we shall take a further step.

The *sixth* entry point is to try to reflect upon the way we have reflected upon the use of mathematics. Reflective knowing must address its own status. This concludes our route towards understanding reflective knowing in classroom practice.

Before coming back to the fifth entry point, let me summarise with a key question, bearing in mind that the entry points are analytically organised (and no more). The list does not indicate any order of sequence and says nothing about how the entry points may appear in educational practice: (1) Have we used the algorithm in the right way? (2) Have we used the right algorithm? (3) Can we rely on the result from using the algorithm? (4) Could we do without formal calculations? (5) How does the actual use of an algorithm, appropriate or not, affect a specific context? (6) Could we have performed the evaluation in a different way? Questions (1) and (2) focus on the mathematical tool, questions (3) and (4) on the relation between the tool and the task, question (5) on the general effect of pursuing the aim with the chosen tool, and (6) looks at the way we have been looking at (1) to (5).

How then does this meaning of reflection relate to the analytical discussion of reflecting on (pointed) mathematical modelling? There seems to exist a gap between the two descriptions of reflection. In the context of school practice, I have not indicated a concise approach to an investigation of the model underlying system development, nor to the effects of the modelling itself. I have formulated the question (5) above in a general form, but it might be challenging to find the meaning of the question in an educational context.

With the fifth entry point we seem definitely to leave the classroom. Does it make sense at all to expose the idea of the formatting power of mathematics in elementary education? Such a perspective was not included in, for instance, the project "Economic Relationships in the World of a Child", but could we imagine the thematic work extended in a way that the question comes to make sense? Let us look at the second sub-project, child benefit, and let us imagine that we try to raise more fundamental questions about the child benefit.[27] The actual payment of child benefits in Denmark has varied according to different sets of rules, so let us try to explore these considerations. Let the amount of money in our fund, F, be fixed. How might we distribute that amount between a number, N, of families? Depending on the age of the students involved in our new project, the question could be specified in different ways. The number of families could be so small that a description of each family could be given: income of the parents, number of children, jobs of the parents, age of children, how they live, etc. The students could then be asked to suggest reasonable ways of distributing the child benefit (cb). One answer could naturally be $cb = F/N$, but at the very least, only families with children should receive child benefit, so we could subtract the number of families without children N_0 and get the new formula $cb = F/(N - N_0)$. However, the amount of money given to each family must naturally depend on the number of children in the family, so if the total number of children is C and the number of children in a specific family f is called C_f, then we get the formula $cb = C_f(F/C)$. But what about the income of the family? And what does it mean if a parent is living alone with the children? etc.

The important question is of course what sort of information about the families is to be used. At one extreme, only very little, at the other, the whole situation of the family could be taken into account, for instance, the distance to work, the importance of having a car for getting to work and for being able to pick up the children from kindergarten. To design a formula in the first case is not so difficult, but to imagine a design in the second becomes nearly impossible. The design of an algorithm presupposes that some simplifications have to be made, and this could be realised by the students during project work. They could gain experience of the difficulties raised by bringing more and more aspects into a formal system. The discussion will easily evolve from what is reasonable and fair to what is possible, relative to the technological tool. Gaining an impression of this sort of development is fundamental to an understanding of the formatting role of mathematics.

A condition for using a formal method is that some degree of simplicity is obtained. Reality has to be rearranged to make it possible to apply mathematics. This is an essential understanding that students must gain, and it raises the questions: Is it necessary to develop a formal method for deciding about the child benefit? Would it be possible to rely on some sort of personal judgement in every case? What is the consequence of this? Would it be possible to use a combination of the two, such that a formal calculation would show what a 'standard' child benefit is, while an individual judgement would modify the formal description once specific circumstances were taken into account? The point is that when it is decided that some sort of formal methods are to be used, then it is also decided that only a limited set of factors can be taken into account. The students could come to realise this as an offshoot of the project – and also that in some cases no alternatives to the process of formatting exist, although alternative ways of formatting exist. Once an algorithm for the distribution of money is fixed, a little part of social life is also fixed. The formatting then becomes a frozen reality for lots of people who have no alternatives but to accept the outcomes of the calculations.

These considerations indicate that it may be possible to illustrate, at a rather basic level in school, what is meant when we say that mathematics is formatting society. It may be possible to show the effects of mathematical modelling. This is important, if it is desirable to provide a linkage between reflective knowing (developed in classroom practice) and democratic competence (including mathemacy as a constituent). I do not say that this approach will necessarily be an educational success; my intention is simply to show what it could mean to illustrate that mathematics has the power of formatting. If this suggestion is feasible then it is possible to develop more examples along the same lines: How could schemes for paying taxes be structured? How could payment and taxes for oil and gas be structured as part of a general ecological policy? etc. Just as literacy can refer to the capacity for finding the political and ideological structures in society, mathemacy could mean the capacity for finding the technological and formal structures built into society. On the other hand, if it all becomes too difficult to find the fifth entry point to reflective knowing in school practice, it may be doubtful whether critical mathematics education is a realistic project. Perhaps to focus on the formatting power of mathematics is not the best way to achieve this.[28]

5. A NOTE ABOUT 'KNOWING'

To identify a technological problem, it is not sufficient to use a formalised language; aims have to be set up in a different language. Technological activity is goal directed, and this directed activity is the object of reflective knowing. With reference to what reflective knowing could mean in the classroom, we have found that parts of such reflections address the mathematical tools for pursuing these aims, parts address the relationships between the tools and the technological aims, parts address the consequence of pursuing these aims with the chosen tools, and finally the reflections themselves have to be put into focus.

Nevertheless, this exposition is too simple because it seems to introduce a logical classification into the concept of knowing. It is not the case that we first need to elaborate a mathematical competence to be able to apply it in pursuing technological aims and, finally, to try to evaluate what has been done. The different types of knowing are integrated in a variety of ways. When students in a daily life situation are preoccupied with financial affairs, they are not simultaneously involved in three different actions. They are buying something and judging whether or not this is reasonable as a single act. When the children in "Economic Relationships in the World of a Child" made suggestions for purchasing equipment for the youth club, all sorts of activities were present in the same situation.

The different types of knowing interact in complex patterns. The students are switching from one activity to another; or more precisely, they are involved in a total but complex activity, which from an analytical point of view contains three different aspects. Instead of maintaining the actual existence of different and self-contained layers, I maintain the existence of a network of interrelationships. I suggest there is a *web of interrelationship between types of knowing*. Reflections may imply changes in technological aims, which may in turn suggest changes in the use of the tools used. It may be necessary to find more appropriate tools, for instance. I do not rely on an additive theory of knowledge development but on a *reactive* theory of knowledge development. This web-analogy explains my interest in the multi-faceted semantics, mentioned in Chapter 5.

However, the web may easily be destroyed, and this is too often done in school mathematics. Reflective elements have been put outside the door of the mathematical classroom and forgotten, and mathematics education has concentrated on the development of a mathematical proficiency. Several epistemologies have served only to emphasise this aspect. The web-structure has been torn apart and mathematical knowledge has been promoted as self-contained. Separation in education destroys the critical potential of mathematics education. Although a scene-setting may be artificial, it still may provide a natural way to develop mathematical, technological and reflective knowing as an integrated structure which it is possible to identify as the competence of mathemacy.

An important step away from the isolation of mathematical knowing has been taken by the applications orientation of mathematics education. Lots of examples of applied mathematics have been developed to illustrate the usefulness of mathematics. This is an important step, but it could easily turn into a stereotype, if it takes the form of informing students about the usefulness of mathematics. The whole complex of mathematical, technological and reflective knowing must find a place in mathematics education. This is a condition for developing the competence of mathemacy. I have used the distinction between the three different types of knowing to draw attention to the fact that reflective knowing has been ignored in the past but it is vital in any attempt to establish a critical mathematics education. Although the distinction is analytical, it has become 'realised' in traditional mathematics education and, therefore, I find it useful. However, the aim is not to maintain the distinction but to make the need for such a distinction superfluous.

CHAPTER 7

"FAMILY SUPPORT IN A MICRO-SOCIETY"

Mathemacy has been analysed in terms of three types of knowing: mathematical, technological and reflective; and reflective knowing was clarified as the competence which could radicalise mathemacy. Thus reflective knowing has been given the task of being a catalyst for critical awareness. Does it make sense in elementary mathematics education to try to involve students in a process of criticising systems established by means of mathematics? As an educational illustration, I shall discuss the project "Family Support in a Micro-Society".

I have described scene-setting as a tool for making it possible for students to see what is being planned to take place in the classroom. Although the content of the scene is factitious, it may provide meaning to questions like: 'Have we made the right calculations?' and 'Could we do without formal calculations?' The educational aim of illustrating formatting by mathematics presupposes a rich context. Furthermore, the problem of transitions between different language games involved in a modelling process cannot be discussed in adequate terms, if no scene-setting takes place. A scene may help to provide a semantic setting for formulating and discussing questions about mathematical modelling and its possible effects. The scene-setting may help to give students possibilities to extract questions relevant for developing reflective knowing.

1. THE STRUCTURE OF THE PROJECT

What scene should we set up in order to make educational sense of questions concerning the formatting power of mathematics? This was a main question in planning the project "Family Support in a Micro-Society", which lasted about two weeks. The teacher, Henning Bødtkjer, and I decided to develop the idea mentioned in Chapter 6: that of trying to describe a micro-society, and in doing so to ask the students to distribute a certain amount of money as a child benefit. However, we decided to talk about 'family support' instead, in order to make the problem more complex. The students involved, 20 in all, were between 14 and 15 years old; the school was Klarup Skole near Aalborg.

Unit 1

The general ideas of child benefit and of financial support for families were discussed. The students were divided into five groups, in most cases two boys and two girls. They had to work in the same groups for the rest of the project. The groups were asked to describe a few imaginary families. They had to include some specific information: the structure of the family, the number of children, their ages, the income of the parents, etc. Besides this, they were welcome to make up whatever other information they found of interest. No outline for the description was suggested. The fictitious families, 24 in all, made up the Micro-Society which was referred to during the rest of the project.[1]

Unit 2

The students had to discuss the standards according to which they wanted to distribute a fixed amount of money: 240,000 DKr. This amount made it possible to provide family support in the Micro-Society at a level similar to that of Denmark in general. Each group could think of themselves as the authorities of a District, and they had to decide about the distribution of family support. Later in the project it would be possible to compare the administration of the support in the different Districts. The students had also to specify what data they needed in order to complete the necessary calculations based on their guidelines for distribution of the money.

Many guidelines were suggested, but a great willingness to make changes was also revealed; one rather common reason was that the suggested principles for the distribution might be too difficult to handle in practice. So already at this initial stage, the students exercised a certain degree of self-censorship.

Unit 3

The students received a nice, typed edition of their descriptions, the 'Family Circle'. Henning Bødtkjer and I considered it important not to distribute the descriptions in the students' original handwriting but to provide the descriptions with the authority of an attractive layout. The students spent some time in reading the stories and to see how their own story appeared in the journal. It was important that they shared the experience. In two of the groups, one of the students read aloud from the Family Circle while the rest listened and made comments. The students were determined to learn about their society. One of the groups also started to take notes for later work.

Henning Bødtkjer and I had added a few extra families to the Micro-Society because it turned out that the students seemed to be fascinated by well-to-do families with good jobs and 'happy' children. Some exceptions existed of course, but Henning and I added a few more.

Unit 4

During the study of the journal, the students realised that it was difficult to keep an overview of the Micro-Society. This made it easy for the teacher to suggest that data could be installed in a computer. The groups therefore had to read carefully through Family Circle in order to pick out the information which was relevant to their principles of distribution. The students' task was to create a database.

In order to install the information in the computer, it was necessary to interpret what was said in the Family Circle. It also became obvious that the descriptions in the journal were insufficient in some respects. The groups had to make decisions about the missing information. The normal way of obtaining this was to ask the student who had written about the family; at least he or she would know. Often the students were rather fast in making a decision: – I see, the age of the child is missing. Anyway, the family lives in number 13, so the age may as well be 13 too!

Unit 5

The groups had to specify an algorithm for distribution. As a consequence the terminology had to be semi-mathematical or semi-algorithmic. In this phase it became obvious that often the general and initially formulated guidelines were insufficient, unclear and ambiguous. If the task was to do calculations, new demands concerning the precision of statements of the guidelines were raised. If the group found the parents' income of relevance, the descriptions had to include, for instance, whether they wanted to adjust the family support to an incremental scale or to use proportionality. The database enabled the students to get new information easily, for example: How many families have a total income below a certain sum of money?[2]

One way to control whether an algorithm was specified by a group or not, was to imagine that they could have somebody help them do their calculations. If they were able to explain to that newcomer how to do the calculations without having to explain all the general principles for the distribution, the group might have specified an algorithm. This definition of an algorithm was understandable and acceptable to the students.

The applications of the algorithms became more imaginative than we had expected, and the shift from a mathematical model to an algorithm did not evolve clearly in the project. One strategy was simply to start with some arbitrary sum of money to be given to a single child, and then to find out how much money was left, and thereafter to find out how to distribute the rest.

Unit 6

The groups had to complete a list of the money received by each of the 24 families in the District. It became interesting to compare the five Districts, and a list, similar to that shown in the next section, was made on the blackboard. The families were taken one by one, and while the description of a family was read aloud by the teacher, the different groups had to write on the blackboard what they had suggested for that family. This made it possible to make a direct comparison. How does the treatment of a family differ in each District? And immediately the question could be raised: Why are they treated so differently? What social policy dominates within the different Districts? Does it reveal a left or a right tendency? Could it be an advantage for the family to move to a neighbouring District? A common comment put forward by the groups in this discussion was: – We agree that a difference exists, but anyway we have made the calculations correctly.

Unit 7

The groups had to write a letter to some of the families stating how much money they would get, and how the District calculated the sum. This task was intended to force the students to try to recapitulate what they in fact had done when they made their calculations.

Unit 8

We planned to return to a general discussion of the principles of the distribution by asking the students to decide upon one final system of distribution. At that stage the discussion would concentrate on the conflict between what was thought of as fair and what was manageable from a technical point of view. This conflict is essential in the discussion of the formatting power of mathematics. But the discussion was not carried out in any great detail.

The more principal aspects of the formatting power of mathematics were touched upon by comparing the distribution of family support with other sorts of social regulations in which the conflict, between what is possible to manage and what is reasonable and fair, can be observed.

2. COMMENTS ON THE PROJECT

Each family of the Micro-Society were given their official number, making it easier to compare the different suggestions from the groups. I quote a few of the descriptions.

Family 3

The Westergaard family lives in a flat in the Vesterbro district in Copenhagen. Laila is 5 years old and plays ice hockey every Monday and Wednesday. She gets 5 DKr pocket money a week, because she is 5 years old. She is saving money for a new ice hockey stick. Esben is 10 years old and is learning to play the saxophone every Tuesday. Each Friday he does folk dancing. Esben gets 10 DKr pocket money a week, because he is 10 years old. Then there is Torkild. He is 14 years old. He does not have any spare time occupations. He is chasing girls. He gets 50 DKr pocket money a week. Finally, there is big sister Pia. She is 21 years old. She has left home and lives together with a guest worker. Pia does not work, so she is on the dole, and the money she gets does not go very far. The father of the children is called Eskild. He is 56 years old and works as a stores clerk. He makes 14,493 DKr a month. The mother is called Christina and is 37 years old. She works as an assistant in a department store. She makes 12,019 DKr a month. Christina has a knitting club together with some friends on Tuesdays. She has just bought a fur coat for 5,000 DKr. She spent her savings in this way. Eskild's old mother died recently, and she left them 20,000 DKr, and with this money they paid their debts. 5,000 DKr were left.

Family 12

Lars is a little boy who lives in the country. His father is a pig breeder and makes 10,000 DKr a month, and his mother works at a school and makes 10,000 DKr a month. They have difficulty in managing after the tax has been paid, and Lars cannot get the horse which he has been wanting for the past two years. Everybody in his class (the seventh) has got a computer, but Lars has got a new front tyre for his 10 year old bike. Everybody gets new clothes all the time, but Lars gets the clothes which his father used in his younger days. Several times his father's farm has been on the point of being sold by order of the court, but each time his father has succeeded in preventing it at the eleventh hour. He inherited some money, but he spent it on renovation of the farm two years ago.

Family 19

The Spore family includes Magnus Spore (father, 40 years old), Grethe Spore (mother, 39 years old), Willy Spore (son, 15 years old) and Anne-Marie (daughter, 14 years old). The father is the vice-president of the largest riding company in Denmark called 'The Winners', and he makes 45,333 DKr a month. The mother has taken over the riding school, named 'Findstrupgaard'. She is working there as a riding instructor, and she has become one of the most famous riding instructors in Scandinavia. Her son Willy and her daughter Anne-Marie win many riding competitions within both dressage and jumping. Willy and Anne-Marie are in the 8th form at the 'Skipper Clements' school. The family lives in the farmhouse of Findstrupgaard. They have 20 horses, 5 wagons, 2 four-wheel driven cars, 1 BMW and a holiday residence in California where they go on vacation each summer. Once they won 6 million DKr tax-free in the numbers game. They also inherited much money from Grethe's father.

Family 24

A short while ago Aase Hansen got a job at a grill bar in a street called Bygaden, and it seems to have become a permanent job. She now makes 11,500 DKr a month. She has two children who spend a considerable amount of time teasing each other. Jacob says that Iben throws her weight about, and Iben says that Jacob eats her liquorice allsorts. Jacob is 12 years old, and Iben is 14 years old. Half a year ago Aase Hansen moved in with Frede Jensen who has a rather large flat. Frede Jensen has never been married. He is an unskilled labourer and now he has a permanent job at the Great Belt crossing. Most of the time Aase, Jacob, and Iben live alone in Frede's flat, but when Frede is on vacation they all enjoy being together. Aase and Frede are not going to get married. They prefer the situation as it is. When Frede is home they share the expenses. Frede makes 21,000 DKr a month. Jacob and Iben do not get any pocket money, except when Frede is at home. Once he gave them 200 DKr each.

The well-to-do families were in the majority in the Micro-Society, but a description of a family of a foreign worker was also included – seven children, and the mother was pregnant again. That family (Family 4) caused some discussion: people from foreign cultures are not always very welcome in Denmark. Family 23 had a handicapped boy; he had to use a wheel-chair and other expensive equipment.

As mentioned, the students had to discuss and find out about their general guidelines for the distribution of the money. Group 4 started by formulating the following principles: (1) The families who earn a lot should receive less than families with a small income. (2) If the children are older than 18 years, the family should not receive anything. (3) If a family has only one child and earns a lot, it should not receive any support, but if it has more than one child, it should receive an amount according to its needs. (4) Families with very young children or with teenagers should receive some extra support. (5) A single parent should receive some extra support. On the basis of these five general principles the group had to find out which data to include in the database.

Group 3 had formulated the following principle for the distribution: Every family should receive the same, independent of their income and independent of the number of children. Without doubt such a principle would save the group from a lot of additional work. Interestingly, however, the group arrived at that principle by a long and intensive discussion. One of the main arguments was that it is also expensive *not* to have children. A couple without children often want to have children, and in many cases this means that the couple become involved in different clinical investigations; and one of the private clinics in Denmark specialising in fertility problems charges people a lot. Later the group changed their radical principle, and became involved in a more complicated modelling process.

Group 2 made calculations which could be used to illustrate the *ad hoc* nature of the used algorithms. First they stated that the total amount of money was 240,000 DKr and that 45 children lived in the society; later, after some extra calculations, the number was changed to 51. From this they calculated the percentage each child could receive – this was 1.961%. Thereafter, they stipulated five intervals of income: above 650,000; 650,000–480,000; 479,999–320,000; 319,999–200,000; 199,999–0. They reviewed their principles: families who earn more than 650,000 DKr will not receive any support, and families who earn less would receive more than families who earn more. Then they graduated the percentages each child should receive depending on the income interval of the family. The five percentages became: 0%, 0.980%, 1.307%, 1.961% and 3.945%. They got these figures by taking the average percentage and modifying the figures according to the income levels. After some calculations the group found that 53,336 DKr were still left. They also wanted to distribute this money. To do so the group used slightly different percentages: 0%, 0.987%, 1.975%, 2.469% and 3.457%. (I do not know how they got these new figures.) After this calculation, 2,897 DKr were still left, but since time had passed the District decided to keep that money. This group was a bit surprised that

TABLE I
The suggested distributions.

Family	Group 1	Group 2	Group 3	Group 4	Group 5
1	13,907.00	20,940.00	13,333.00	9,880.00	10,077.00
2	8,814.00	8,381.00	13,333.00	6,000.00	5,043.00
3	18,314.00	18,069.00	19,999.00	12,000.00	7,543.00
4	46,128.00	42,161.00	46,665.00	19,600.00	35,077.00
5	3,000.00	0.00	0.00	0.00	2,543.00
6	8,814.00	18,069.00	6,666.00	11,800.00	15,077.00
7	18,314.00	12,571.00	19,999.00	8,800.00	7,543.00
8	32,221.00	20,052.00	33,332.00	14,600.00	12,543.00
9	9,500.00	6,023.00	6,666.00	4,000.00	5,077.00
10	8,814.00	8,381.00	13,333.00	8,000.00	5,043.00
11	6,000.00	0.00	0.00	0.00	5,043.00
12	4,707.00	6,023.00	6,666.00	5,000.00	5,077.00
13	8,814.00	20,940.00	6,666.00	6,000.00	10,077.00
14	0.00	0.00	0.00	0.00	0.00
15	3,000.00	0.00	0.00	0.00	2,543.00
16	0.00	5,755.00	0.00	1,800.00	5,043.00
17	6,000.00	0.00	0.00	0.00	0.00
18	8,814.00	12,046.00	13,333.00	8,332.00	5,043.00
19	6,000.00	5,755.00	0.00	4,000.00	5,043.00
20	6,000.00	5,755.00	0.00	1,800.00	5,043.00
21	4,407.00	12,046.00	6,666.00	2,800.00	2,543.00
22	6,000.00	0.00	0.00	1,800.00	5,043.00
23	8,814.00	5,755.00	13,333.00	5,800.00	5,043.00
24	8,814.00	8,381.00	13,333.00	1,000.00	10,077.00
Total	245,196.00	237,103.00	233,323.00	133,012.00	171,184.00

their first calculations did not come out exactly. They had thought of their original procedure, not as an *ad hoc* way of making the calculations, but as an exact one. The list of the suggested distributions from the five Districts are shown in Table 1.

It is interesting to compare the suggestions. There seems to be some general agreement. Family 4 got the most in all Districts, this was the family of the foreign worker. It was also agreed that Family 14 should not receive anything, in fact that 'family' consisted of a single old man going to celebrate his 90th birthday together with his children and grandchildren. In general the suggestions for the different groups did not add up to the total 240,000 DKr. In Group 4 the difference is rather high, but as they said: – Our District always has to save money if possible.

The students were finally asked whether they would have distributed the money differently or not, if they did not have to use mathematics. Some of the comments were: – Well, I think we would have distributed the money differently, if we had not had any model. I think we would have taken a few more things into account. I think that we still would have taken into consideration income, number of children, and the children's age, as we did in the model. However, perhaps also the total assets of the family. Perhaps we would have made a more fine-graded scale. For instance, I would have given Family 13 a bit more than the 8,814 DKr than we actually did. (From Group 1.) – Our distribution is based on the model we made. I do not think we would have distributed the money differently. I wouldn't. I think the money has been distributed in a fair way ... In fact, I think that only parents who have no real chance of bringing up their children should receive money. The others do not need the money. (From Group 3.) – In one way, I think, I would have done something different, if I had a complete family description. Something more could have been taken into account, for instance whether divorced parents received other forms of support. (From Group 5.)

3. REFLECTIVE KNOWING IN THE PROJECT

In a modelling process transitions between different language games occur, and this kind of transition can be found in the project "Family Supply in a Micro-Society". The Family Circle constitutes 'reality' according to the scene-setting. The first linguistic transition took place when the database was established. The students had to go through the descriptions of the families, and from the jumble of information they had to pick out what was relevant in order to set up a database. This step is similar to that of system development in general. Such a development presupposes interpretations of reality, and at least two types of interpretation took place in creating the students' database. First, it had to be decided what information was relevant, and this decision depended on the selected criteria for the distribution. For instance, it had to be decided whether the ages of the children were relevant – and what about the ages of the parents? Naturally a solution was to incorporate too much information in the database, but even that depends on an interpretation. (In the description of one of the families it was mentioned that one of the children had freckles, and this information was not included in any of the databases.) Further, the information was stated in colloquial form, and sometimes an interpretation was needed in order to distil information in the form of numbers. For instance, how could data be installed about a family in which the parents are separated, and the husband lives together with his secretary, while the children live most of the time with their mother, and the father is paying an insufficient amount of money to the mother?

Both the decision about the scope of the relevant information and the crystallising of information into 'data' presuppose a certain perspective. Gathering data is a complex activity. In the general description of the process of system development it was mentioned that such development is based on a more or less well-founded theoretical framework and on a more-or-less biased set of interests. Data do not mirror any objective reality but represent a condensed perspective. To establish data as the basis of a system is a creative process. It is an act carried out by means of the systemic language. This conception is repeated in the project, and the students had to go through a process of system development of the same nature as an authentic process of system development.

A different linguistic transition takes place when a mathematical algorithm of distribution is specified. In the project it became difficult to distinguish between a mathematisation and developing an algorithm. In fact a step seems to be taken directly from the developed system, into an algorithm. Once the students had selected their set of data they tried to describe how to distribute the money before trying to make a general mathematical model. Originally I delineated the different activities in the mathematical modelling process in an order of time but, as mentioned, such an order is misleading. Also it turned out during the project work that developing an algorithm for the distribution influenced the students' guidelines for the distribution. They implicitly changed the principles of distribution according to what they thought was manageable. The two transitions, that of system development and of developing an algorithm, became the principle transitions in the project.

Different fundamental questions are connected with formatting. One has to do with the disharmony between what is seen as worthwhile and what is workable. This disharmony is also found generally when solutions to a social problem are pursued by technological means. The conflict may take the following form: We find the best way of solving a problem to be so-and-so, however the only technically feasible solution is so-and-so. The structure of that conflict indicates how a discussion referring to visions and ideals becomes substituted by a discussion referring to technical possibilities. This substitution often accompanies mathematical formattings. The different language games involved in the modelling process do not leave much room for ethical discourse.

Whether or not it is possible to create a situation in which the students come to realise such a conflict is an educational problem. However, during the development of the algorithms for the distribution of support, the students came close to problems of that nature: – We cannot do it in that way, it takes too much time! If we try to take that factor into consideration, we shall never get finished! During the project the conflict was sometimes solved just simply by the students 'forgetting' about their original guidelines. Also they wanted to finish their work. During the initial groupwork some principles were set up for the 'ideal' distribution but because of the complexity of (even) the micro-society, these ideals were impossible to achieve. When mathematics is applied, simplifications are introduced. During the process of system development, visions of an ideal action become modified into manageable calculations. What is technically possible becomes a substitute for what may have been outlined as the optimum. The discussion of norms and values can easily be eliminated by the discussion of technical possibilities. This objectivisation is a general aspect of mathematical formatting, and the students involved in "Family Support in a Micro-Society" were introduced to a discussion of such phenomena.

These comments outlining features of the modelling process have been somewhat negative. The comments have implicitly been closely related to an interpretation of the relationship between natural language and formal language: The interpretation shows that the step from a rich natural language into a rigid formal grammar is a step which ignores important possibilities for gaining insights. The terms used have been those such as 'simplification' and 'making things manageable'. As set up in this section, mathematical formatting is seen primarily as problematic, and reflections become concerned with limitations and difficulties. The modelling process has, so to say, been described as a variant of an *inverse alchemy*.

Clearly, it may be possible to find other interpretations which would give credibility to supporters of formalisation and not accept the rosy picture of natural language painted by 'ordinary language philosophers'. By using mathematics, sometimes it is possible to find solutions not possible to identify without mathematics. Modelling has to do with a transition between language games, and the grammar of mathematics also provides possibilities not attainable by natural language.[3] It is necessary not to think of the transition from verbal criteria to mathematics as always leading to simplification and conciseness. During system development something may be lost, but we may gain something differ-

136   CHAPTER 7

ent. If, for instance, the distribution of family support were to take place without any sort of algorithm, then the distribution may be subjected to more arbitrary guidelines. To develop a mathematical algorithm also implies a way of discussing and obtaining universal guidelines. Mathematics may provide discussion with a stronger set of arguments, and therefore 'golden rules' may also be a result of mathematical modelling.

4. UNDERSTANDING 'FORMATTING'

The illustration of the formatting power of mathematics is an essential aspect of "Family Support in a Micro-Society". Mathematics 'freezes' the algorithm for what is to be done, and in this sense the (implicit) mathematical model becomes a guideline for action. The model manifests norms for distribution. It has been emphasised that the thesis about the formatting power of mathematics does not suggest that mathematics is the only agent in social development; we are concerned with an interplay of different forces, one of which is mathematics. It has also been emphasised that the thesis does not imply that mathematics, as such, possesses a formatting power. But in what sense, then, could mathematics alter society? What is the meaning of 'mathematics' in such formulations? It is obvious that it does not make much sense to substitute 'mathematics' by 'mathematicians'. No conceivable interpretation can be given to 'mathematicians are formatting our society'. The project "Family Support in a Micro-Society" was developed to provide an educational illustration of mathematical formattings, but it also indicates that 'formatting' needs further consideration.

Formatting was originally described, in abstract terms, as a path leading from thinking abstractions to realised abstractions. This terminology provides an idealistic formulation of the potential of mathematics and its effects. However, the fact that I have rephrased the realisation of abstractions as being concerned with transitions between different language games transforms the original conception. 'Language games' draws anthropological thinking into linguistic philosophy. By making statements like 'mathematics is formatting our society', we readily construct an abstract entity; a reification of mathematics takes place. It now seems relevant to talk about the mathematical community as a socially constituted group which produces results that have a possible use in a different context. Such an approach is more in accordance with what actually was illustrated in the project. 'Formatting' also refers to a social context, and therefore the formulation 'mathematics is formatting our society' becomes a shorthand formulation of: The scientific community of mathematicians is a socially constituted group socialised into a paradigm of mathematical practice which can be defined by different 'coordinates': a language which can be described as semi-formal; a set of meta-mathematical views; a set of accepted questions; a set of accepted methods; and a set of accepted statements.[4] And this paradigm produces the tools by means of which technological aims can be pursued in a new and effective way. By providing analytical equipment and tools for technological action, the output of mathematical practice comes to play an unique role in our society.

I do not point to any interpretation of mathematics as being the 'prime mover' of society. We have to recognise an interplay, and we always have to take into consideration different social forces, one being economical interests another being technological development. I shall not try to establish a determinism claiming that economic forces are dominating, nor a technological determinism claiming that the inner trend in technology specifies social development. I only maintain the existence of an interplay of different forces in which outputs of the paradigm of mathematical practice play a role. If we recapitulate what happened in the project, we find that mathematics as a technological tool interacts with a policy for setting up a system for distribution, but the tool is not subordinate to the guidelines for the system development that takes place, nor does it completely dominate the original aims. An interaction takes place. Let this be a meaning of 'mathematics is formatting our society'.[5]

An aim of reflections is to get hold of the nature of mathematical formattings, and hence help to define mathemacy as a general critical competence. As part of critical education, literacy refers to the competence of making interpretations of social phenomena, and according to the thesis of linguistic relativism such phenomena are also constituted by language. If mathematics has a formatting power, this means social phenomena are partly structured by a formal language, and therefore mathemacy may have a role similar to literacy. I hope to have illustrated this by "Family Support in a Micro-Society", and for this reason I find the project useful for ascribing meaning to critical mathematics education.

The setting up of a scene is important because a discussion of a system created by mathematics does not make much sense if the modelling is not embedded in a situation which demands questions, and which is perceived by the students as worth investigating. The discussion of the formatting power of mathematics cannot progress unless we have a natural-language description of a situation in which the formattings takes place. Formattings cannot be evaluated from within a formal language. If the background scene set up is subtracted from the project, "Family Support in a Micro-Society", the whole educational process could be split up into a number of small exercises. It is not difficult to see that all elementary mathematical competencies which could be developed during the project could have been achieved by means of a normal sequence of mathematical exercises. Some of the technological competencies may also have been developed by involving students in such a sequence of exercises. The point, however, is that decomposition into exercises would destroy the possibility for developing critical reflections about the whole activity. Reflections, and therefore the development of mathemacy, presuppose a situation that provides a discussion with a semantical content. This makes scene-setting important in critical mathematics education.

My point is not that, given suitable scene-setting, reflection will flourish among students. My description does not claim to expose a story of educational success. To embark on an investigation of what actually may happen between students and how they really act in the conceptually identified conditions for developing reflection, is quite a different empirical task.[6] Here my point is simply to relate the meaning of scene-setting to conditions for reflections. A scene provides a semantical ladder, but it does not push anybody up that ladder. If the students were not involved in a process which made it possible to discuss what it was they were engaged in, they would not have had the opportunity to perceive the conflict between what is technically feasible and what is preferable between one set of principles and another.

5. A NOTE ABOUT CHALLENGING QUESTIONS

It seems obvious that during the project "Family Support in a Micro-Society" some questions could have been addressed more directly to the students. The students set up general guidelines for the distribution of family support, but later, when involved in the task of making the actual distribution, they sometimes 'forgot' about their guidelines. They became *absorbed* in the technical task of making the distribution.[7]

This phenomenon of absorption is of general importance. A difference exists between the discourse which results in guidelines for some aspects of management, and the discourse used when the actual management is carried out. In the project we saw the difference between developing a system and developing an algorithm for realising this system. The students became involved in two different types of discussion depending on either being 'politicians' setting up guidelines or 'technicians' making things work. The discussion among the students turned from what is desirable to what is manageable. This is a most critical step. We could see the transition from the first to the second question as a transition between two different language games (two different discourses), the first being directed towards goals and values, the second towards technological possibilities. The questions moved from 'What do we want to do?' to 'What are we in fact able to do?'.

This is a most fundamental transition. It cannot be avoided when mathematical modelling plays a role. As we have already noted, transitions between different incommensurable language games are essential parts of the mathematical modelling process. But it is possible to prevent the whole investigation from becoming absorbed by the technical perspective. To reflect upon technological questions presupposes a 'return of the absorbed'. The technological discourse has to be embedded in the broader discourse in which evaluation may take place. Technological absorption on its own is dangerous because it conceals uncertainties. If the results of a technological device, produced in an absorbed manner, are to be evaluated, it is important that a new transition takes place. It is not possible to use the medium of technological discourse if we are to evaluate the results of a technological device. The languages of mathematics and algorithms are insufficient for the evaluation of such devices. These languages produce tunnel vision which makes it difficult to see the importance of reflective knowing.

When absorbed in a technological question, it becomes difficult to move from one form of discourse into another. However, it would have been possible during the project to confront the students with their absorption in a much more direct way. After making their suggestions for the distribution, the different groups could have been confronted with their original guidelines. An essential part of the educational strategy could have been to release the students from their technological absorption. I see this as being an essential aspect of critical education. The final distribution made by the different groups could have been contrasted with each other, although they also revealed a strange similarity. No groups had taken into account the specific problems of Family 23. In the description we learnt that the family has a handicapped child, who needed a lot of expensive equipment. Why did the groups not take that into consideration? As one of the students commented, they would have taken such an aspect into account had they not had to base the distribution on calculations but on intuition. But the point was not pursued further. What we have here is an epistemic obstacle, and this obstacle is similar to the general obstacle of how to conduct an evaluation of devices constructed within a technological absorption.

If we, in an educational situation, have to take the step from making calculations and applying mathematics to reflecting upon what has been done, *challenging questions* seem to have a special role to play. Maybe it is necessary to provoke proceeding from a basis of technological actions in the direction of reflective knowing. Therefore, challenging questions acquire a specific epistemic meaning.[8] No continuum of knowledge development leads from having mathematical abilities to an understanding of the use of mathematics. The importance of challenging questions is an essential idea to learn from the distribution project. Naturally, it always has to be kept in mind that we are acting out a scene. The students know that whatever they make of the distribution is of no consequence. They are acting out a fictitious 'reality'. Anyway, the epistemological idea which I suggest we note is that the development of reflective knowing is of dialogical nature. More than one voice is needed if normative perspectives are to be pursued.[9]

CHAPTER 8

"OUR COMMUNITY"

If mathematics is a predominant ingredient of today's technology, then it becomes important to highlight this as part of a mathematics education which tries to establish a means for organising and reorganising "interpretations of social institutions, traditions and proposals for political reforms."[1] Critical mathematics education, then, means an education which tries to criticise authentic, real-life applications of mathematics. But even if we have succeeded in providing an educational interpretation of mathematical formattings and, in so doing, have given the term 'reflective knowing' a sound educational meaning, we have yet to discuss the *scope* of the reflections. To concentrate on the formatting power of mathematics is perhaps too severe a limitation making the conception of 'reflection' too narrow.

I have described 'mathemacy' as analogous to the concept of 'critique of ideology'. Some features of our conception of social life are so deep rooted in traditions and norms that they become invisible and difficult to question, and the task of a critique of ideology becomes to put such hidden conformations back on the agenda. In this sense a critique of ideology becomes an intellectual enterprise, and to some extent this 'intellectualisation' is duplicated by the idea that reflective knowing has to address questions about mathematical modelling. Critique becomes after-thoughts, or after-modelling-thoughts, and reflection becomes a process of coming to understand the assumptions and effects of the application of some formal techniques. But we must also question whether we have touched upon the genuine *nature* of reflective knowing. The concept of 'reflections' may have been developed in a too analytical and academic direction.

Mathemacy, with reflective knowing at its centre, has been specified as the competence which defines critical mathematics education. An uncertainty about the scope and the nature of reflective knowing, therefore, indicates that we cannot claim to have found a conclusive account of critical mathematics education. Besides, we find a conceptual gap between the loosely held notion of democratic competence and the notion of reflections about formal methods. No robust philosophical connection has been established between the concept of reflective knowing and 'life' in a democracy. We have uncovered a connection by means of the thesis about the formatting power of mathematics and the assumption that living in a democracy presupposes an awareness of the structuring principles of society: but a more straightforward connecting route might exist.[2]

Perhaps we have met with another complication as well. It need not be the case that a complicated competence, such as mathemacy, can find space for developing in a curriculum chopped into different subjects. Perhaps an essential condition for developing mathemacy is that it is not confined within a curriculum of mathematics education. Perhaps the reflective component of mathemacy presupposes interdisciplinarity, implying that, literally speaking, the concept of 'critical mathematics education' contains a contradiction. While mathematics does not include a language about mathematics, reflections do presuppose such a language, and therefore reflections demand interdisciplinarity. I shall however try to give a short-term 'solution' to this problem. By 'critical mathematics education' I do not have in mind any attempt to outline a different curriculum in mathematics, but critical mathematics education may refer to a form of educational practice in which mathemacy may develop. This practice may be interdisciplinary, problem-oriented, project organised, etc. However, the nature of the interdisciplinarity in which critical mathematics education can grow has to be discussed, and I shall try to relate the discussion to a new example. I shall not concentrate on further elaboration of 'mathemacy' but will try in a more direct way to relate interdisciplinary educational practice, including mathematics, to an interpretation of democratic life in a local community.

1. THE STRUCTURE OF THE PROJECT

The village Hinnerup is in the centre of the Hinnerup District, and the local authorities and the town council are located here. The project "Our Community" took place in a comprehensive school in the village Hinnerup. The intention of the project was to give the students an idea of some of the conditions and possibilities, not only for living and working in a local community, but also for influencing its social and political life. Can school prepare students to participate as citizens in a democracy, not only as receivers of the 'output' from the authorities but also providing an 'input' to the community? By raising this question, the project led to a close involvement with the notion of democracy, and because the District is rather small (about 10,000 inhabitants in total) the original Rousseauan conception of direct democracy came to make some sense. The project tried to show that it also has an educational meaning.

Fifteen students were involved in the project which was planned, organised and carried out by the two teachers, Jørgen Boll and Jørgen Vognsen. The students were about 16 years old. They all were in the tenth form, which is the last form before they have to leave secondary school. Therefore, the students had to decide what sorts of jobs they wanted to be educated for or whether they would try to attend upper secondary school. In Denmark it is quite normal for students to practise different sorts of jobs for a short period of time to get an idea of what it is like to have a job and to have the opportunity to formulate (realistic) hopes and plans for the future. Such a one-week 'trainee service' was included in the project "Our Community", but with the restriction that all students had to find a 'job' in the Hinnerup District, which naturally had already been agreed to by the authorities of the District. First the students were introduced to the general idea of the project, and then followed a two week period fully concentrated on project work. During the first week the students had their trainee service, while in the second they were working in their classroom. I summarise the project in the following units.

Unit 1

The intention of the project was presented and discussed with the students. They came to know about the trainee service and the project work to follow by means of questions dealing with the District. Also some of the issues which had caused heated political discussion in Hinnerup were introduced. One issue had to do with the position of a traffic-relief road, a second concerned the heating supply of the district, and a third, the establishment of a youth club.

Unit 2

The students were introduced to the 'job news' posted on the noticeboard in the classroom. One of the notices ran as follows:

Kindergarten Assistant
The Red Cross Kindergarten has an opening for an Assistant (38 hours a week) for participation in practical and educational work with the children as well as other activities. Applicants who want an education in the socio-educational field will be preferred. Working hours: 7 a.m. to 3 p.m. or 9 a.m. to 5 p.m. Salary and terms of employment according to existing collective agreements. For further information, please contact Elin Filtenborg, tel. 86-911233, extension 612, Monday from 1 p.m. to 2 p.m. Please address the application

including relevant documentation to the District of Hinnerup, The Mayor's Office, Nørregade 1, DK–8382 Hinnerup.

The notices, which were drawn up by the local authorities in collaboration with the teachers, all had the normal official wording, and the vacant jobs, 24 in total, covered the whole range of possibilities in the District: occupational therapist, porter, teacher, environmental supervisor, unskilled labourer, librarian, technical assistant, etc. The students had to find out which jobs they wanted to apply for, and the teachers helped them to make appropriate applications. Some of the jobs were more attractive that others, especially the jobs in the Technical Department.

Unit 3

When the District had received the applications, the students were given appointments for job interviews. The interviews were carried out in a realistic manner (the interviewers were used to doing the sort of job interviews involved). Some of the students could not get the job they had applied for simply because more than one had applied for the same job, so they were assigned other jobs. The process of getting a job was conceived of as vital by the students, knowing that in a few years' time they could be put in a similar, but real, situation.

Unit 4 (One week of trainee service)

Besides participating in the job for a one week period the students had to answer some general questions, formulated by the teachers, about their workplace. The questions had been presented in advance to the workplace by the teachers so that the staff were prepared for what the students were going to ask about. The questions concerned: The students' general experience during the week of work practice; the nature of the services which the institution provides for the District; the structure of the decision-making in the institution (who takes the decisions and where does the money come from?); and the changes which are expected to take place in the institution in the future. Besides answering these questions, the students had to make a short report about their work experience. Generally, the different workplaces were very open to the newcomers, and one student came to participate in a regular staff meeting dealing with some of the problems of the institution.

Unit 5 (Monday morning)

After the week of trainee service, there followed a week fully concentrated on selected problems of Hinnerup. During this week the students worked in their usual classroom. As a general introduction, a chief clerk from the local authorities was invited to give information about the structure of the Hinnerup District, especially about the relationships between the politicians, who (in principle) take the decisions, and about local government employees who prepare the groundwork for decision-making and who also have to carry out the decisions. The lecture had the nature of an introduction to new members of the local staff – and it was also made realistic in the sense that it became a bit boring.

A second theme that morning was raised by a person from the local Social Security Office. He described some of the opportunities this part of the Civil Service could offer. In Denmark we have a huge unemployment problem among young people, so what was described could all too soon become part of the students' reality.

Unit 6 (Monday afternoon)

The students had received different reports and leaflets with information about the Hinnerup District concerning town planning, population forecasting, etc. They were also given a loose-leaf file with additional information as well as with different questions and tasks to deal with during the rest of the week. Some of the tasks were carried out as group work, but in some cases the students worked in pairs or individually. The questions and tasks had to do with the following: (1) finishing the report of the trainee service period, (2) finding out about the population forecasting for Hinnerup, (3) making some suggestions with respect to the economic situation of the local Music School, (4) finding out about plans for different sections of the community concerning house building, and, finally, (5) making up a 'dictionary' containing the unknown words which the students had come across in reading the different reports and information about Hinnerup.

It was not possible for all the students to be working at the same task at the same time; for instance, working out calculations for the economy of the Music School presupposed access to a computer, but only a few were available.

Unit 7 (Tuesday morning)

This morning was organised as group-work concentrating on finding out about population forecasting for Hinnerup. The students had gathered statistics specifying the number of persons in the different age groups. They had to look at the development of the group of fifteen to sixteen years old, to illustrate the forecast by a diagram, and to try to explain the tendency in the forecast. The students themselves were that age, so the forecast would also provide information about the number of tenth forms in the future. (It became obvious that some decrease would take place, and that not every school in Hinnerup District could expect to have a tenth form in the future.) Further, the students had to compare the developments of different age groups. More general implications of the forecast concerned the programme of house building, building of schools, the capacity of kindergartens and the capacity of rest homes. The students had to compare their conclusions with the official descriptions of the programmes for the District.

Unit 8 (Tuesday afternoon)

This afternoon a lecture was given by a civil engineer employed at the Department of Town Planning. The main subject of this session concerned suggestions for a new traffic-relief road connecting two different areas of Hinnerup which are divided by the railway tracks. The City Centre is close to the railway, while the residential areas are placed North and South of the Centre. How could the North and the South be united without destroying, for instance, a large part of a forest area close to the City Centre or other areas of recreational value? This question had created a major discussion in Hinnerup. Different interest groups confronted each other: What was the most satisfactory solution from a technical point of view? What was environmentally more preferable? And the cheapest solution? These questions also had to be discussed in terms of different local interests which influenced the evaluation of the suggested solutions.

The students were able to participate in the discussion with the civil engineer because that specific subject had been dealt with as part of the introduction to the project, and because the students, during this Tuesday morning, had acquired additional information about conditions for town planning. A small-scale physical model of one of the suggested solutions was placed in the classroom, and this made it possible for the students to get a better idea of what the implications of a solution could be. The encompassing topic for this afternoon lecture was: Why is it necessary to have town-planning in a community?

Unit 9 (Wednesday morning)

This session was organised as group-work concerned with the Music School in Hinnerup. This school gives lessons to students interested in practising music. (The existence of the Music School also shows that it is possible to realise a local initiative if sufficient support is mobilised.) The Music School is organised as a private school, meaning that the expenses are not covered by the State Government but are paid by the parents of the students who attend lessons. However, in many other towns in Denmark the local authorities support the music schools financially.

The students had got the necessary information about the economy of the Music School, and before the project "Our Community" started they had learned how to manage the sorts of computational problems raised on a computer by means of spreadsheets. This meant that the task of making a budget and calculating the consequences of different possibilities was manageable. The students were working in groups, and the results of the calculations were organised as recommendations for the finances of the Music School and related to support from the District as well as to the structure of the payment. For instance, the prices for smaller children could be different from the prices for older ones, and the prices could depend on the type of activity – it is cheaper to organise a chorus than to teach one child how to play the piano. Further, a reduction could be introduced if more that one child from a family attended the Music School. The proposed budgets were drawn up in quite realistic terms and the proposals were, in fact, handed over to the head of the Music School.

The students' investigations, however, continued. It was calculated what it would mean, in terms of increased taxes, if the authorities decided to support the Music School, and it was discussed whether local organisations other than the Music School could also argue for the right to receive financial support. In this way, emphasis was placed on the fact that it does not make sense in a democracy to take care of an isolated case without discussing whether a particular decision has general consequences or not. It was also estimated what more general financial support to different cultural activities in Hinnerup would mean for the level of taxes. In this way the students got some idea of possible reasons for politicians sometimes having to say 'no' to spending extra money.

Unit 10 (Wednesday afternoon)

This afternoon's session was introduced by a local politician who talked about how to solve two (quite different) common tasks in the community: the heating supply of the district and the Civil Defence. The latter topic became the most fascinating because it turned out that the school was designated to become an emergency hospital in case of war. In the cellar beneath the school, where the students previously had never been allowed to go, they saw the facilities available for the hospital. It was obvious that the materials in stock were totally insufficient, in fact it was quite ridiculous to call it an emergency hospital – it was just a few beds and some blankets.

In a very straightforward way, this experience raised questions such as: How much do we have to spend on an entirely hypothetical situation, a war breaking out around Denmark, knowing that most likely (and hopefully) we shall never have to use that money? What do we think about the miserable emergency hospital stock beneath the classroom? How can we compare expenses for the hospital with expenses to be used, for instance, for the Music School?

Unit 11 (Thursday morning)

This morning was organised as group-work. The students had to continue the different tasks which they were given at the beginning of the week (Unit 6, Monday afternoon).

Unit 12 (Thursday afternoon)

The Mayor visited the class. A particular subject for this afternoon session was the possibility of establishing a club house for young people in Hinnerup. It was obvious that the students had gained some ideas during the week relevant to such a discussion; they had more ideas about the structure of the decision-making in a local community and about the economic possibilities and limitations.

Unit 13 (Friday)

This last day of the project was fully occupied by the students finishing different tasks and also by their presentations to the rest of the class of their various results.

2. COMMENTS ON THE PROJECT

It is obvious that the project "Our Community" was strongly supported by the local authorities in Hinnerup, primarily because the project had been so carefully prepared by Jørgen Boll and Jørgen Vognsen. It was also most convenient that the project was set in a small community where people knew each other, for example the head of the Music School was also a teacher at Hinnerup School, and the Mayor had himself previously been a teacher.

The project "Family Support in a Micro-Society" described earlier was concentrated on the formatting power of mathematics. However, the project "Our Community" raises the question of the extent to which a competence, comprising reflective knowing in our original interpretation, has, in fact, to play a crucial role in developing a democratic competence. "Our Community" suggests that I might have described too narrow a concept of reflections by strongly identifying schooling as a preparation for a democratic life. Added to this, mathematics education can be part of such an enterprise without having to take the roundabout way of reflecting on mathematical modelling processes.

It is obvious that the intention of "Our Community" was not to make education an uncritical preparation for obedient citizenship. It becomes a gross simplification to interpret 'education for democracy' merely as an introduction for students to the basic values of a democratic state. An introduction to democracy also presupposes that students become familiar with ways of seeing democracy as involving negotiation. Learning for democracy could also mean learning how to interact with authorities, and in this case, 'learning by doing' makes sense. Therefore, education for democracy must include a critical look at the background of decision-making in a concrete context.

"Our Community" was based on a carefully elaborated scene set up by the teachers, and it was easy for the students to identify with the scene. We have recognised different types of scene-setting: the realistic one as exemplified by "Economic Relationships in the World of a Child" and the present project "Our Community"; the imaginative one as exemplified by "Golfparken"; and the activity-based one in which a variety of tasks are specified for the students to accomplish as exemplified by the workshops set up in "Constructions". A scene-setting may provide connections between 'reality', which in the case of "Our Community" was not very remote from the classroom, and educational activities. Nevertheless, I find scene-setting of particular importance even in this case. It would have been possible to teach all the different facts about the Hinnerup District to the students in normal classroom situations, but that would have determined quite a different education, not necessarily having anything to do with 'education for democracy'. The artificiality of scene-setting is also a way of making connections with reality. The fact that the students had to apply for 'jobs' at the Hinnerup District and to prepare for the job interviews was 'pretend reality', but nevertheless this made the project more realistic. Scene-setting is a way of communicating a meaning of an educational process. It provides a language about what is done during educational practice.

In Chapter 3, I analysed 'democracy' in relation to the question of delegation of sovereignty, and we come to focus on the term 'democratic competence', seen as the opportunity for (and the ability of) the majority to control the people in charge. Further, this competence was analysed as containing an important element of mathemacy, encompassing reflective knowing, because it was found important to the participants in a democracy to be able to identify the formatting power of mathematics. However, with the project "Our Community" we return towards the original interpretation of democracy as outlined by Rousseau, which I declared, perhaps a bit too hastily, to be of minor educational importance. The present project indicates that mathematics education can be integrated as part of democratic education in a more immediate way. The students come to investigate some of the background information for the decisions taken by the authorities of the Hinnerup District Council, as for instance concerning the programme for house-building. They come to investigate the possibilities for alternative decision-making, as for instance with respect to the Music school. In both cases they have to use their mathematical background. Further, they are not only observers of democratic routines but become involved themselves. They had the opportunity not only to listen to the Mayor but also to contribute in a face-to-face discussion with him. Naturally, such close contact with the authorities is not a necessary condition for 'education for democracy', but I think this situation could symbolise what is meant by an education not only striving to make students reactive but also active in a democracy.[3]

The present project shows that *Mündigkeit* can be given a specific interpretation, such as the students being able to participate in political discussions taking place in a local community. It is not necessary to confine this concept to the level of philosophical generality and close to a critique of ideology. The concept of *Mündigkeit* had dwelt in the conceptual heaven of Critical Theory since Adorno put it there, but "Our Community" has brought it back into educational practice, showing that the concept has a simple and direct interpretation.

Clearly, it is also possible to find potential for investigating the formatting power of mathematics in "Our Community". Such an investigation could be based upon the analysis of the economy of the Music School. It would be possible to continue an investigation of payments and to find out in what way different payment policies display different attitudes towards families, children, economic background, elitism, etc. A system of payment could be adjusted to the parents' income making it possible for everyone interested in music to participate, or it could be adjusted to the ability of individual students, making it possible for the most able students to receive the most intensive lessons from the most highly qualified teachers – just to mention two extremes. It is not difficult to see that if "Our Community" was extended in such a way, it could come to cover many of the features of mathematical formattings.

The overall question, however, is whether "Our Community" *has to be* developed further to illustrate the formatting power of mathematics, if it is to fit into a conception of mathematics education as part of a democratic force and to become part of critical mathematics education. Is the essence of critical mathematics education linked with identifying the formatting power of mathematics and with developing reflective knowing in relation to applications of formal methods in society?

This question makes it necessary to return to the concept of critique. This concept does not only refer to a competence but also to the attitude or disposition of an individual. A person need not be critical even if he or she knows whatever is necessary to carry out a critique. A critical awareness must also be presupposed, and making room for such an awareness can be seen as an educational task. This, I shall take also to be part of critical mathematics education. It is not sufficient to look just for the proper object for a critique, it is important as well to look for personal reasons for carrying out a critique.[4] We cannot limit critical mathematics education to being only a criticism of the applications of mathematics. Democracy refers to a competence of participation and to an awareness, and I find an important aspect of critical education to be the opportunity for students to be able to interact with people in more powerful positions. A democratic competence is not just a reactive but also an active competence, and I find this illustrated by "Our Community".

In "Our Community" a variety of activities were carried out outside the classroom. It does not make sense to substitute a traditional mathematics education with a critical one, leaving the structure of the timetable untouched. Some of the boundaries around the mathematical classroom must be broken down for some periods – but I do not argue that a permanent 'breaking down' has to take place. In "Our Community" the eradication of the boundaries around the subject took place over a fairly long period, and perhaps this is what is essential. The traditional teaching of mathematics cannot be substituted with critical mathematics education without attacking the regular timetable structure. Critical mathematics education demands a degree of interdisciplinarity which presupposes structural flexibility.

Two examples of mathematics were involved in "Our Community": one had to do with graphical representation to analyse the population forecast in Hinnerup, and the second had to do with the budget-making for the Music School. Common to both was the fact that the calculations provided rationales for action. In this way both examples came to exhibit the formatting power of mathematics. It is not possible to avoid action. It is necessary to have a house-building programme, and it is necessary to have a budget for the Music School. An essential experience for the students was that the forecast for the development of the population in the Hinnerup District was not more realistic, even if mathematics was involved. Still something unpredictable could happen.

3. REFLECTIVE KNOWING – AN OPEN CONCEPT

Reflections manifest the competence of grasping and evaluating the effects of a technological enterprise. Reflections can concern technological tasks on both a large and small scale, and I have mentioned six entry points to reflective knowing as developed in a classroom setting. The entry points were identified by the questions: Have we used the algorithm in the right way? Have we used the right algorithm? Can we rely on the result from the algorithm? Could we do without formal calculations? How does the actual use of the algorithm, appropriate or not, affect a specific context? Could we have performed the evaluation in a different way? These questions clearly can be generalised, but still I find that reflection can mean much more. To me "Our Community" is essential, because it opens up the conception of 'reflective knowing'. The project "Family Support in a Micro-Society" was developed with the original interpretation of reflective knowing in mind, with the primary intention of illustrating the formatting power of mathematics. "Our Community" has made obvious to me that reflection has various dimensions, and these dimensions were already present in "Family Support in a Micro-Society".

Scene-setting facilitates reflection; this is shown by the questions indicating entry points to reflecting. How is it possible to ask in any meaningful way 'Can we rely on the calculations from the algorithm?' and 'Can we do without formal calculations?' without scene-setting? The presentation of mathematics as a sequence of exercises makes a parody of reflection. A scene-setting is a way of providing a semantics which makes it possible, perhaps even important, to reflect upon the results of some calculations. However, a scene-setting also opens up possibilities for many sorts of reflection other than those indicated by the six small entry points. I will mention some possibilities integrated in "Family Support in a Micro-Society" as an example.

– The students were writing descriptions of the families: – What do I include? Does the family I am going to describe look like my own family? Do they live a better family life? In which family would I prefer to live? The teacher asks us to include some descriptions of not-so-rich families. Why does he ask this? What does 'poor family' in fact mean?

What are my classmates writing about? Are they writing something which has to do with a family I know? Could it be my family? Why did the teacher ask Peter to change some of the names in the story he is writing together with Lisa? Is my description as good a Peter's and Lisa's? Why am I writing my story alone, while most of the others are working in pairs?

– The students set up some guidelines for distributing family support. They discussed those guidelines, and to set them up they had to choose priorities. They were also involved in a sort of self-censorship. They modified some of the guidelines which they originally had set up: – What do I put into the database? It seems fascinating to type in information. I like using the computer. If only the others would find out what to write, I just like the computers.

– The students were writing short letters to some of the families to explain about the support the family has received: – The families are invented, so why do we have to write these letters? The teacher wants us to do so, so let us write some letters. Well, how do we explain what we in fact did? This seems a bit difficult, and why write about it because we have already done it? We are finished. I do not understand why the teachers ask us to do this. The others seem to write some very short letters, I think I will do the same.

– The results of the distributions were written on the blackboard while the teacher read aloud the descriptions of the families: – Did we really take everything about that family into consideration? Strange, but I seem to have forgotten many things about the different families. That was a strange description, who wrote that story? I cannot remember anybody having written about a woman having a job at a grill bar. Could it be the teacher, or perhaps that fellow from the university, who has invented that story? There seems to be a great difference between the suggestions. And how is it possible that the group over there always gives less that the rest of us? They cannot have used all the 240,000 DKr. They must have been rather lazy when doing the project.

– The teacher wants us to discuss the result, but what does he mean? What are we going to discuss? I do not understand. But Lisa seems to have something to say. She always does. Is it true that a family could receive so much money from the District? I am almost sure that we do not receive that much money at home. How much money do we in fact receive? How much do the fictitious families in fact receive? They cannot receive so much money per month – but is it per year? Or what? I cannot remember that anybody mentioned this, but they must have talked about it sometime. I do not think I will ask.

Many possibilities exist for reflecting upon making the distributions for family support as well as reflecting upon the educational process itself. Clearly, in all types of education students think about what they are doing and about what is happening in the classroom. The point is that in the case where a scene-setting has taken place, those reflections may have a better chance of entering the public domain of the classroom. A scene may include reflection as part of the agenda.

The multi-faceted semantics brought into the classroom by scene-setting makes reflection an open concept. It does not make sense to restrict reflection to having only to do with thinking about mathematical modelling. But even if it were an advantage to open up a concept, this step also contains a danger. If reflection comes to include more and more, it not only becomes a richer concept in terms of breadth but also more shallow. To open up a concept not only means to make it more applicable, it simultaneously makes it less useful. This is the danger of the step I have suggested. Besides, I have a new difficulty. Originally we have set out to define critical competence in terms of mathemacy including reflective knowing, but what becomes of the content of 'mathemacy' if 'reflection' is given a more open structure?

If we compare the two projects "Family Support in a Micro-Society" and "Our Community", we find a tendency to move away from mathematics. This may be seen as a problem to (traditional) mathematics educators, but for the more philosophical task of clarifying educational ideas, the tendency indicates only the possibility that some of the essential aspects in critical mathematics education do not concern mathematics. However, we shall take a fresh look at a different example which may indicate that mathematics, on its own, may contain a critical potential.

CHAPTER 9

"ENERGY"

Two general aims were set up for the project "Energy". First, it should be simple for the students to see and to understand what they were going to do, and therefore what was happening in the project should be related to the life experiences of the students. The second aim was that an exploration of some of the topics of the project should possess an 'exemplary' value. As discussed in Chapter 4, the thesis of exemplarity says that it is possible to develop a general understanding by concentrating on (becoming absorbed in) a specific example. The intention of the project was to develop and analyse situations from daily life in such a way that attention could be drawn to some global problems concerning the use and supply of energy.

The project was divided into three sub-themes. The first had to do with energy and food: how much energy is contained in a certain meal, and how much is used by performing a specific task (cycling)? The second sub-theme had to do with farming and two questions were addressed: How much energy is used in order to produce barley in a field the size of one hectare, and how much energy can be expected to be reclaimed from the harvested barley? And: When pigs are fed with barley, how do the input–output figures for energy then look? The third sub-theme had to do with electricity: How much electricity is used in each of the student's homes, and is it possible to reduce the use of electricity?[1]

The students involved in the project, 20 in all, were about 15 years old. The form was non-streamed, and this means that a great diversity existed concerning the 'ability' of the students. The intention of the project was also to provide a situation in which both 'able' and 'less able' students could cooperate and still meet different challenges. The project was organised at Klarup Skole, and the teacher who planned and carried out the project was Henning Bødtkjer; Andreas Reinholt from Aalborg Teacher Training College was co-ordinator.

1. THE STRUCTURE OF THE PROJECT

The project can be described in different units. As previously, each unit could be constituted by a single lesson as well as by several.

Unit 1

The students had to get to school without having eaten breakfast at home. At school each student had to weigh carefully how much breakfast he or she got: bread, butter, cheese, etc. (Obviously the project had been introduced before that morning.) Then each student calculated how much 'energy supply' the breakfast contained, and they gained an impression of what that expression could mean. They had collected statistics giving information about the energy supply contained in different sorts of foods, and this information was entered into computers. For one of the students the calculations resulted in the following: bread: 59 g + 29.5 g (which is equivalent to 1,000 kJ), cheese: 20 g + 20 g (which equates with 642 kJ), butter: 9 g + 4 g + 5 g (which equates with 558 kJ), etc. In this case the total energy supply was 3,050 kJ.

The calculations resulted in comparisons between the students: – The two of us had nearly the same amount of food, but I seem to have received much more energy supply than you. – But that has to do with your use of sugar. Look, how many kJ sugar contains, and I did not cover my cornflakes with sugar! The comparisons resulted in some remarks about the difference between quantity and quality of energy supply, and also the idea of quality of food was touched upon.[2]

No preliminary calculations had been completed by the teacher to indicate any approximate answers, so to both the students and the teacher the results provided new information. Without any further comparison, however, the results of the calculations would be too abstract. But, a perspective could be provided if the students got some idea of 'using energy'. How much energy is used by performing a specific task?

Unit 2

This unit consisted in a bike trip. In this way the students got a tangible idea of the meaning of using energy. The problem set was how to express these phenomena by more specific calculations, but the students knew that the teacher had some formulas in mind. All the students had brought their bikes to school, and they had to take a trip the distance of which was known. Different types of bikes were brought to school; some had racers with all sorts of equipment. One student said: – This morning my tire was flat but my aunt lives just around the corner, I borrowed her bike. It was harder to ride an old bike than a modern racer. How was this to be reflected in the calculations? The formulas which were going to be used also contained the parameter 'frontal area of the cyclist', but how were they to calculate this area?[3] The teacher used a video camera, and all the students had to ride directly towards the camera to be filmed.

Unit 3

This unit engaged the students in long and complicated calculations to show how much energy they had used on the bike trip. To do this calculation, the specific distance (8,130 meters), the time taken (for one of the students it was 1,380 seconds), the 'frontal area of the cyclist' and the type of the bike had to be known. Three different formulas could be used to determine the 'bike resistance' r which depends on the type of bike, the velocity v, and the 'frontal area' a:[4]

$$\text{Normal bike:} \quad r = 1.1av^2 + 7$$

$$\text{Sports bike:} \quad r = 1.0av^2 + 6$$

$$\text{Racer:} \quad r = 0.7av^2 + 5$$

The frontal area was estimated by each student individually by using a video image of themselves riding towards the camera. To get the right scale each student had attached a little piece of cardboard on which were drawn a few squares, one dm^2 each. The top of the cardboard was fixed to the student's sweatshirt by two safety pins. This meant that the cardboard took a vertical position independent of whether the student was sitting in an upright position or bent forward when riding the bike towards the camera. The whole video print was divided by drawing a lattice of squares over it, and that made it possible to count the number of squares needed to cover the whole picture of the person. One student estimated her frontal area to be 0.41 m^2. It was possible to compare these results and it became obvious that the frontal area depends on the person's position when cycling.

The formulas determining the bike resistance needed quite a bit of explaining, but from this basis it was possible to calculate the energy used by applying the following formulas (which also needed explaining):

$$\text{Cycling power:} \quad c = 3.6rv$$

$$\text{The cyclist's use of energy:} \quad e = (4c + 400)t.[5]$$

For the student using 1,380 seconds for the trip (using a racer) the energy used was estimated to be about 640 kJ. Some of the students were surprised that so little energy had been used when they compared the results with the energy supply received from the breakfast. For a majority of the students the formulas were not too difficult to apply, but it was a problem to understand the ideas behind them. However, the main intention was not to explain some physical and physiological background but to give the students the opportunity to apply formulas, and to provide them with the experience of what it could mean to use energy.

Unit 4

Some time was spent obtaining a more general interpretation of the results. The results had been difficult to predict but once arrived at, they seemed to make sense. Different comments were raised: – If I used 640 kJ for the trip, how far would it be possible for me to cycle using the energy supply from my breakfast completely? For one student one breakfast seems to equal five trips on the bike. Does this result make sense? To agree about this was also a way of evaluating the results. – What would have happened if I had cycled longer? The cardinal points were concluded in input–output terminology, and the expression energy-account was also used.

Graphs were drawn to illustrate the relationships between different types of consumption of energy, and as a minor example, it was calculated how much energy a newspaper deliverer, running up and down staircases, was supposed to use. It was calculated to be 4,020 kJ per hour, which far exceeded the energy used when cycling. It was also calculated how long a time it would take to consume the energy supply from a bag of chips – it surprised the students that it would take such long time. The students also experimented by putting together different dishes to find out what hypothetical meals would contain. This was not too complicated because the information was available on computers.

Unit 5

Introduced the next sub-theme: use of energy in farming, and the students prepared for what to look for during the excursion to a farm. The problem was introduced as a variant of the input–output question. Farming results in an energy output in the form of the production of food, but how much energy does the farmer have to use for that production? What do the input–output figures look like in farming? It was obvious that the farmer needs to prepare his fields to get an output and that he has to supply some energy to produce an output. It was also estimated by the students and the teacher that the step from preparing a field for sowing to harvesting would be a 'good' business, but that the step from feeding pigs with corn to producing bacon was more dubious (from the perspective of energy consumption and production, but not necessarily from a financial point of view).

What should one look for if one has to find out about input–output figures? What should be looked for when arriving at the farm? Is it at all possible to make formal estimations of what is happening in farming? A difference between the input–output question for breakfast-cycling and that of farming was that in the first case, a mathematical model was available while in the latter, no mathematical model was known. The students and the teacher had to imagine what sorts of calculations would provide a result. The students came up with suggestions, for instance it became obvious that the width of the different tools used in preparing the field must be important. Then they went to visit the farm to try to find out what it was all about. In this case neither the teacher nor the students had any idea about the result. What do the input–output figures look like in farming? Nobody could know anything before some calculations were done.

Unit 6

When the students arrived at the farm they started measuring out an area of one hectare to get a specific idea of the size of the field they were going to use as the unit for calculations. The task of measuring out a field the size of one hectare was surprisingly difficult for the students, but finally the field had a post in each corner. The students and the teacher walked together around the field and tried also to stand in the middle. – Bigger than a soccer ground! Then back to the question: How much energy is used to produce barley in that area?

The farmer gave them an interview in which he explained the different steps in preparing the field, including: harrowing, sowing, rolling, spraying, harvesting and ploughing. The students also had to take notice of how many times the field had to be worked over by the different tools. Later they looked at the tools used and measured how wide they were to calculate later the amount of petrol needed in preparing the field. When they knew the width of the tool, they could calculate the distance the tractor had to drive.

Later the students had a look at the pigs, and experienced the smell and repulsive noise in the stable. The pigs were bought by the farmer when they were still small, and their weight was only about 25 kg. – How much do they eat? How long do they stay on the farm? What is their average weight when they are sent to the bacon factory? The students needed all this information to calculate the input–output figures for energy in pig-breeding. The students saw the mountain of barley in the barn, and in fact the feeding of the pigs took place in a fully automatic way. Barley was sucked up from the mountain and then given in the right doses to the pigs, adjusted to their weights.

Unit 7

Back at the school the students started to calculate the use of energy necessary in the production of the one hectare field of barley. They had specified three types of input: seed, fertiliser (including spraying), and petrol for the tractor. They knew the distance the tractor had to travel: It was estimated that for a field of one hectare the distance added up to 35 km per year, they knew the amount of seed which had to be used in sowing, and they had received information about the fertiliser and the spraying.[6] Then followed complicated calculations, and they were all waiting for the result. It was very exciting, because nobody knew anything, but everybody had some ideas.

Some students became lost in the calculations, so the teacher had to supervise and help a lot. But it was essential that everybody understood the general question guiding the calculations, and that the aim of the calculations was not lost sight of. In one of the students' calculations the use of energy added up to 13,090,000 kJ. The resulting energy, contained in the harvested barley, was calculated to be 81,650,000 kJ, so a short summary of the result showed that the farmer gets back the energy six times when he produces barley – the sun is his helper.[7]

Unit 8

Then the production of pork was put into focus. This phase of the calculations was also organised as group-work, and it was initiated by the general question: What do you think is important information in order to estimate the amount of energy used in pork production? The simplified answer was that they had to know the weight of the pigs when they arrived at the farm and when they left, the time the pigs stayed in the farm, and how much food they ate in relation to their weight. From a handbook of 'Planning in Farming' the students obtained information about the growth of pigs and how much they ate. Based on these statistics, it was possible to calculate how much energy supply in the form of food a pig needs from the time it arrives at the farm until it is sent to the bacon factory. One student calculated it to be 2,635,000 kJ.

The next step was to calculate how much energy is accumulated in an average pig when turned into bacon. The weight of the pig is essential, but naturally also how much runs to waste. The result was 496,000 kJ. This means that approximately a fifth of the energy put into that sort of animal production is recouped. From an ecological point of view, animal production is extremely expensive.[8]

Unit 9

These results were put into more general terms, and naturally it had to be kept in mind that the results were highly approximate. Were the calculations reliable? Was the result inaccurate, meaning that more accurate information of the type already gathered would improve the result, or had some essential parameters been forgotten, meaning that no conclusions could be drawn? This part of the project took place as a general classroom discussion. The results were also discussed with the farmer, who thought they were basically correct. Put into a global perspective, the students learned about the basic conditions for the production of vegetables and of meat.

It was also discussed whether it was possible to improve the input–output figure for pork production, and therefore the space in the pigsty was debated. The smaller the space the faster the pigs will increase their weight, but such an approach also has a limitation: Is it legitimate to breed pigs with only input–output figures in mind?

Another point was touched upon: What does the energy account look like in different parts of the world? It would be interesting to compare statistics from the U.S.A and, for instance, China. If personpower as a factor of energy supply is not taken into account, the production of corn in the U.S.A. is perhaps much more expensive and inefficient than in developing countries. The students were very involved in these discussions, and were proud that their calculations raised perspectives of a global nature.

Unit 10

This unit consisted of a parents' evening. The parents knew about the energy project and were now introduced to its third phase: Use of electricity. The main question was: Would it be possible to save energy in the students' homes, if more was known about the consumption of energy of the different electric devices: vacuum cleaner, refrigerator, deep freezer, hair dryer, etc.? Would it be possible to find some strategies for savings?

Unit 11

The students had got the following three bits of information: the recent prices of electricity; the condition for the production of electricity; and the fact that only a limited amount of coal is left on the earth. Therefore it makes sense to try to reduce the use of electricity. The basis of this part of the project became a cooperative effort between each home and the school.

Unit 12

The students were assigned the job of investigating the use of energy at home. Their first task was to make a diagram of their house showing the different electrical installations and devices to get an overview. From the school each student had borrowed an instrument for measuring the use of energy of each of the electrical devices, and the main question was: Which one is the greatest user of electricity? It became something of a detective job to identify the 'big user', and different consequences followed. In one home some of the light-bulbs were changed to low energy bulbs; in another the deep freezer was moved to the garage. Often the reduction took place by changing the temperature in the deep freezer, too often the temperature was much lower that necessary. It was possible for the students to borrow a number of low energy bulbs from the school, and this made it possible to change all the bulbs at home to see if any change could be noticed.

The students measured the use of energy at home for a normal one-week period, then a saving-week was introduced. Was it possible to reduce the use of electricity at home, based on the knowledge of the energy used by the different electrical devices? The use of electricity for the two weeks were compared – some reduction had taken place.

Unit 13

In this unit the students ran a programme: 'Save electricity!' at the school. They knew about the use of electricity for a normal one week period. They had acquired that information from the porter of the school. Then they introduced a saving-week at the school by identifying possibilities for saving electricity. Different sorts of posters and pamphlets were produced, but the students came to realise how difficult it was to take the message about energy saving out of their own classroom to others. The porter of the school could not say there was any noticeable reduction of the use of electricity during the saving-week.

Unit 14

Back in the classroom things were put into a global perspective: What happens if the use of electricity is not reduced? What sort of consequences can be expected? This summary also became the introduction to a new project: construction of a windmill.

2. COMMENTS ON THE PROJECT

The formulas used for calculating the 'bike resistance' seemed hard to understand, but it was possible for the students to discuss these formulas. They had a general idea of what the formulas meant, because it was possible to relate them to physical experiences. As mentioned, different formulas were used, depending on the type of bike. It was not too difficult to see that the constant in the formula for the racer had to be the lowest one. The racer is quite simply easier to ride forward: – Well, perhaps the aunt's bike should have a constant higher that 7, but perhaps this constant could be lowered if the bike was oiled better. It was also obvious to the students that the frontal area had a role to play, as well as the speed. It was more surprising that the speed had to be squared. However, this feature of the formula also equated with something in the students' experience: it becomes harder and harder to increase the speed, the faster you are cycling.

The formulas used in the input–output models for breakfast-cycling were looked up in sources from sports medicine; they were simplifications of specific formulas actually used when bike-resistance was calculated. The students knew this, and this provided the formulas with an authority normally not ascribed to formulas in mathematics teaching – often traditional mathematical exercises use invented formulas. The goal was to enable the students to understand the basic ideas in the formulas – not to provoke criticism of the actual content of the formulas, this was expected to be too difficult. However, the students' attitudes towards these 'reliable' formulas were quite different than towards invented formulas. In fact, invented formulas from textbooks seem to retain a much higher authority in traditional mathematics teaching. What they express may be right or wrong, but not much attention is paid to such possibilities because the primary task of mathematics education is different. An exercise has only to do with the application of the given formula, and the result is considered to be right if the formulas are applied correctly. Whether the result as such is useful or not, reliable or not, is irrelevant. As an example: in a textbook two formulas state the growth of stalactites and stalagmites as a function of time. Whether or not the calculations give a reliable answer for the growth is completely irrelevant to the students, and the formulas used can preserve their authority in silence. In contrast to this, the formulas for bike-resistance were discussed, and their plausibility was evaluated, even when the students knew them to be grounded in research. In this case the students did not show much respect for authority, and it became urgent to find out whether or not the formulas had an intuitive reliability.

This had to do with the subjective importance of the results.

Different types of terminology were used during the project. Mathematical expressions were developed by means of which the calculations were conducted. It was rather easy for the 'more able' students to make the calculations, while the 'less able' needed help – and in fact they received substantial support and advice from their friends. A second type of terminology was developed when the discussion of these formulas took place: What is the general idea of the formulas? And, surprisingly, this talk about formulas became less abstract than might have been expected and, in fact, understandable to students not able to do the calculations on their own. The reason seems to be that they had gained a personal feeling of what the formulas might mean. This was surprising, because investigating formulas in general terms is normally thought of as an abstract task. But the students were able to interpret the general ideas of the formulas. What does it mean that the formula contains the parameters 'frontal area' and 'speed', and not other parameters? What does it mean that the product of the frontal area and the speed (squared) has to be multiplied by a factor, depending on the type of bike? What does it mean that a constant has to be added? (And what does the constant signify when the speed is zero?) It makes sense to distinguish between the ability to do formal calculations and the ability to understand what the formulas symbolise. The ability to carry out calculations seems not to be a necessary condition for understanding the 'jobs of formulas' where students have some informal experience of what the formulas are expressing. Thirdly, we have the terminology concerning the purpose of the calculations, and the cardinal point here is that the general explanation given by the teacher was made in terms understandable to (perhaps) all the students. The input–output terminology is quite commonplace and it is possible to express it in a comprehensible way, because the students already had a specific experience of 'input' and 'output' as well as of their relationships: – How long are we able to cycle before needing a sandwich? A similar phenomenon could be found concerning the input–output models in farming. Even if several of the students were unable to perform the calculations on their own, they were still able to understand what the calculations were about – and anxious to know about the final results.

This phenomenon can be related to the distinction previously made between mathematical, technological and reflective knowing. I do not maintain that this distinction reflects the original distinction, but it accentuates the possibility that mathematical knowing is not a necessary condition for the development of technological competence and, further, that a fully developed technological competence is not a necessary condition for reflective knowing to develop. It is possible for students to comprehend the general content of the formulas without actually being able to perform the calculations, and it is possible to understand the idea of an input–output model without grasping the specific nature of the formulas. The understanding of what takes place in the process of mathematics education is not a linear addition of understandings, the basis of which concerns mathematics. Understanding for the students is more like an awareness of the web of relationships referred to earlier.

The way in which the project "Energy" was organised meant that the students themselves could not become directors of the educational process, but during the educational process some *vantage points* were established from which it was possible for them to 'see' what had been achieved and what was going to be done during the next phase of the project. The direction of the educational process and the purpose of the different tasks did not have to be expressed in mathematical or technological terminology, but it could be expressed in terms related to the daily life experiences of the students or to experiences they had all developed as part of the project. When a vantage point is established, it is not an educational catastrophe if students do not understand the details of the calculations to follow. They still have a meta-conception of what they are doing.[9] It was essential to the project "Energy" that the figures of the 'energy-account' were provided with meaning. The students had a physical experience of energy supply as well as of use of energy, and therefore the idea of an input–output model was given a very tangible meaning. (It was noted that one had to cycle almost two hours at a good speed to use up the energy from half a bag of chips.) The meaning was not only important for an interpretation of the calculations but essential as well for establishing vantage points from which the students could get an overview of the project. The summaries of the particular sub-themes did not stratify the students according to 'ability' (although this might have been the case when it came to the calculations).

To calculate the use of energy in farming for the production of barley, the relevant information was the width of the machines used and the number of times the field has to be worked over, and not to have a specific perception of the size of a one hectare field. So what was the function of actually measuring out the one hectare field? As mentioned, it was rather difficult for the students to find out how to outline a one hectare field. The students were a bit bewildered but finally they found out how to do the measuring, and they decided that their field should have the dimension: 200 m times 50 m. From a mathematical point of view the actual measuring of one hectare, as well as walking around the field, were quite unnecessary, but we have to take into consideration that mathematical knowing is not all that is at stake in mathematics education. The measuring of the field was essential for providing a point of reference for later discussion. When it came to a summary, i.e. when it was necessary to have a vantage point in the educational process, this one hectare field became cardinal. Later on, several references were made to this piece of land, and it was possible to provide summaries with a well-established concreteness: – That the calculated input–output figures show the proportion one to six means that the farmer can harvest barley containing about six times the energy he has to use when harvesting, sowing, ploughing and whatever else has to be done to that area we walked around. Naturally, a statement concerning the input–output figures does not contain more essential information whether or not a reference is made to a particular field. The difference is that the teacher can make a reference to a situation in which the students have participated. The terminology used for explaining the general ideas of the calculations is provided with a new meaning. So the activity of measuring can be interpreted as a way of enriching the semantics for the discussion of the educational process. Measuring out the field provides a vantage point with meaning. If these students are not to lose interest, it is important that vantage points in the educational process involve the students as participants, then it is possible to make references like: 'When we were biking . . . ' and 'When we walked around the field . . . '.

One of the main purposes of scene-setting is the production of vantage points. Vantage points become hills in the semantical landscape of project work. From a vantage point it becomes possible to get an overview of what the project is about, because it helps to provide a meta-language about the different tasks in the educational process. Vantage points become semantical conditions for communication between students and teacher, not only about the strict educational content but also about the perspectives of this content, and about what is done and what has to be done.

3. FORMAL LANGUAGE VERSUS NATURAL LANGUAGE

The project "Energy" draws attention to some of the limitations of the previous descriptions of the role of mathematics in society. It has been emphasised that mathematics has a remarkable effect on society, and the expression 'mathematics is formatting society' has been coined. This has produced the idea that coming to see the difficulties in making transitions between the different language games involved in the modelling process, and to see how mathematics creates patterns for technological actions and design, are most essential in an educational process in which 'mathemacy' is related to 'literacy'.

The notion of 'inverse alchemy' has been mentioned, but an underlying assumption has now to be questioned. Is critical mathematics education primarily to be seen as a reaction to the inverse alchemy of mathematical modelling of social affairs? This has been an assumption in developing the project "Family Support in a Micro-Society". The project "Energy" shows that this focus is not the only one possible – as long as we do not exclude "Energy" as an example of critical mathematics education. "Our Community" has already made an important addition to our notion of critical education, and "Energy" also makes an addition.

'Mathematical archaeology' designates an attempt to identify and name the actual use of mathematics behind the polished surface of technology. However, the project "Energy" did not include a mathematical archaeology, although the project could have been extended in that direction. It is not difficult to find possibilities to reflect about the reliability of formulas – for instance concerning the formulas of bike-resistance – but the point is that these reflections do not define the value of the project "Energy" (although it definitely has some value). The project approaches mathematics in a different role to that of prescribing reality, and this different role also has to be discussed in relation to critical mathematics education. This draws our attention once more towards the philosophical interpretation of the relationship between formal language (being systemic, mathematical or algorithmic) and natural language.

Two different positions have been mentioned: What could be termed the Russell–Carnap notion, that natural language makes room for confusion and ambiguity while a formal language can become the underlying and uniting structure of adequate descriptions of reality; and the Ryle–Austin notion which says that formal languages are gross simplifications of limited aspects of natural language making formalisations highly dubious as a means of description, except perhaps for some very restricted areas. The Russell–Carnap notion and the Ryle–Austin notion represent two different positions concerning relationships between types of language and 'reality'; and a comprehension of mathematical modelling reflects a position somewhere between these two extremes.

'Reality' however is a flimsy concept. To allow it refer to non-linguistic phenomena which can be an object for description cannot sustain much examination, if we do not rely on a Picture Theory, as outlined in the *Tractatus*. But as already mentioned, a Picture Theory produces too simple an interpretation of language. Instead we must take into consideration the complexity of the functions and effects of the use of language. The descriptive aspect is only a single aspect, and descriptions and interventions by means of language become mixed together in a speech act theory, anticipated by the notion of the formatting power of mathematics. The discussion of the Russell–Carnap notion versus the Ryle–Austin notion cannot be reduced to a comparison of descriptive potentials but must take into account what can be done by the different types of language games.

Nevertheless, let me briefly concentrate on the descriptive aspect, now to be seen as a one-dimensional shadow of the total function of language. Formal language and natural language become two different means of description. They cannot simply be translated into each other, and it is unlikely that they encompass the same features of reality. Part of reality can be described by mathematics,[10] and in some cases natural language descriptions may be adequate. Clearly, it may be possible to see an overlap between the range of mathematical descriptions and natural language descriptions, but we cannot expect coincidence. Some features of reality lie outside the reach of mathematics but inside the possibility of natural language descriptions and *vice versa*.[11] In some cases formal language descriptions can be the most suitable, they have a potential not shared by natural language descriptions and this means that in some cases, formal descriptions have potential to express features impossible to reveal in any other way. This expresses a strong thesis about the utility of mathematics, but in my interpretation this is accompanied by a similar thesis stating that in other cases natural language descriptions have a power which exceeds the potentials of mathematics.[12]

I do not assume that everything can be described by an adequate language. I do not assume that language, being formal or natural, can get to the boundaries of 'reality'. This means that the thesis of the *Tractatus*, i.e. that the limits of language are identical with the limits of the world, is negated. None of our descriptive tools can be expected to be exhaustive. Nevertheless, a language can express too much. 'Fata morgana' (illusion) in (scientific) descriptions is always a possibility. Let us first look at natural language descriptions. Such descriptions always run the risk of embodying more than may actually be the case; the observer may be said to observe reality and simultaneously some of the (ideological) features built into the language. Natural language is – in this interpretation – seen as not only a means for communication and as a tool for description, expression, actions, etc., but also as a catch-all for a variety of assumptions built into the grammar of the language, and these features may be projected into reality by the act of description itself. When observing and describing we also observe our language. Language creates its own template of reality. If we do not take into consideration the phenomenon of language projection and that natural language is a socially constructed set of conditions for interpretation, we shall never gain an awareness of the phenomenon that language may cause fata morgana. However, similar remarks can be addressed to the use of a formal language, because this language also builds upon a grammar which sets out conditions for observation and interpretation. Put in a simple way used in attacks on logical positivism: by attempting to apply a formal language as a descriptive tool, reality is transformed into a set of facts suited precisely for that type of descriptive tool. Formal language becomes just as problematic as natural language when it is not assumed that the logical structure of language is similar to 'the logical structure of the world', and this thesis – also formulated by Wittgenstein in the *Tractatus* – is difficult to maintain. Also formal language produces the fata morgana phenomenon, and if we do not have any means of adjustment, we may be tempted to ascribe to reality the logical features emerging from the grammar of that type of language.

The endpoint is that I do not find it possible to make any general conclusion about complementarity between a Russell–Carnap notion and a Ryle–Austin notion concerning the relationship between formal and natural language. The complementarity has to be discussed with reference to particular situations.

The types of illusion, with respect to meaning caused by natural language and by formal language, cannot naturally be expected to be identical, although we can fear an overlap. The existence of an overlap implies that even if qualitative and quantitative descriptions can be used in a sort of complementary critique, they cannot be expected to rule out all forms of fata morgana. The very idea of description links up to the possibility of such illusions. When we try to describe by means of a language we may always observe features in the description which belongs to the descriptive tool but without being able to identify those features. Nevertheless, the qualitative and quantitative means of description are important in the complementary critique of each other.

When we try to understand what language might be doing it is essential to be aware that we may see too much or too little by means of the language used. (As emphasised, description is just one dimension of the jobs of language.) In some cases we may expect mathematics to exercise a power not found in natural language. Up to now the developed perspective of critical mathematics education has focused on the formatting power of mathematics, which relates to the phenomenon of fata morgana: mathematics may describe relationships which are not actual, but reality can be changed, and fata morgana becomes real by an act of formatting. Reality is changed to fit the mathematical description of a 'might-be'. It has been emphasised that mathematics education must be able to identify this phenomenon and from this perspective, the theory of reflective knowing has been developed. But this is not a sufficient background for critical mathematics education.

If mathematics in some cases has a unique descriptive power not identical with the descriptive power of natural language, some features of reality are only accessible by means of mathematics, perhaps even some critical features.[13] We might be able to do something by means of mathematics which we would be unable to do with natural language alone. A mathematical competence may mean an empowerment, also in terms of critical education. And this means that critical mathematics education cannot be developed solely through the thesis about the formatting power of mathematics. To me the project "Energy" indicates not only that the social role of mathematics has to be discussed further but also that the notion of mathemacy can be expanded.[14]

4. COMMENTS ON MATHEMACY

If we look at the way the input–output problem of farming was discussed and analysed in the project "Energy", it becomes obvious that we are not dealing with an example of formatting – at least not in any immediate interpretation of formatting. Mathematics was used to express some correlations which cannot be simply identified in any other way. Something can be seen and done by a formal language not possible to see and do by means of a natural language. The results from the investigations of the students could not have been expressed and argued in the same specific way without calculations. It could have been suggested that some problems in pork production exist, but it is not possible to express any scale of the problem, if we do not perform calculations. Here we find an example of a unique power of mathematics made visible in an educational context. Naturally, it is necessary to emphasise that formal language does not support any independent argument for what is seen. The importance of mathematics lies in a different place. When mathematics is used in a description, as the one in question, it becomes possible to analyse hypothetical situations also. For instance, now it makes sense to ask how the input–output situation may change if the use of petrol is changed. Would it have some implications for the use of energy, if the field was prepared in a different way? etc.

To talk about fata morgana seems to presuppose that the essential function of language is to describe 'actuality', but fata morganas may be useful. Mathematics might describe hypothetical situations and make it possible to investigate 'might-be' situations. When mathematics is used, we can be involved in hypothetical reasoning about reality. And this aspect of mathematics can be seen as part of the "Energy" project. This introduces, in fact, a new conception of mathematics. It might be interpreted as an important source for building up hypothetical situations making it possible to make thought experiments. When a bridge construction is planned mathematics might be used for carrying out experiments concerned with the stability of the bridge without actually first having to build the bridge. We can experiment with an economic policy without actually having to carry out political reforms (this has been the essence of constructing the SMEC model, for instance). This aspect of thought experiments could be found in the project "Energy", if further investigations had been carried out. What would be the result if the field were prepared in a different way? How much petrol would have been used in that case? Would it be possible to change the input–output figures of energy in a radical way? I see the creation of possibilities for such an investigation as the essence of the application of mathematics.

Sometimes it might be impossible to compare a hypothetical situation with 'reality', sometimes it does not even make sense to make comparisons with 'reality'. In the case of bridge building, the hypothetical situations might have some resemblance with reality, but how do we compare an investigation using SMEC with economic reality? The economic reality is constructed by the technological means – by means of which it is analysed. However, understanding hypothetical situations compatible with an actual situation, creates possibilities for action. Mathematics might provide an overview of situations which is impossible to investigate if only natural language is at hand. And the range of accessible alternatives to an actual situation becomes a determinant for actions. Sometimes it seems impossible to investigate sets of hypothetical situations with a resemblance to 'reality', but because we often are forced to act on only a partial insight into alternatives, a formal language approach to hypothetical situations can also produce irrationalities in actions. Formalisation and irrationalism can also go hand in hand.[15] But still, in some situations irrationalism is all we have, and mathematics becomes a means for real problem solving.

As previously stressed, mathemacy is composed of a mathematical, a technological and a reflective competence, and reflective knowing provides mathemacy with a radical sharpness. However, in "Energy" we saw that mathematical competence became a vehicle for an important area of understanding. The insights gained by the students during the project "Energy" could not have been developed without mathematics, and this emphasises that mathemacy is an integrated competence. It makes sense to see mathemacy and literacy as competencies which provide means for interpreting social affairs. The use of mathematics made it possible for the students to interpret what they observed about farming. It would be impossible to outline any input–output figures without mathematics. The use of mathematics provided an opportunity for seeing causes for the observed phenomena. If it is questioned why the amount of energy is increased by a factor of six when producing barley and reduced by a factor five when vegetable food is transformed in pig breeding, then the students have an explanation for those figures as a result of their participation in collecting the relevant data. They cannot be sure that their justification is sound, but they have learned something about the nature of a justification. The students are placed in a stronger position, if they have seen that the result does not vary fundamentally even if some of the data are not correct. Naturally, these observations always have to be accompanied by a more fundamental doubt: Perhaps some important features have been left out. It is not possible to ensure by means of formal analysis that no blind spots occur. However, in the project "Energy" mathematics describes relationships which may have similarities with real structures in farming and these relationships become open to discussion because of their mathematical formulation. The descriptions turn into hypotheses which can be improved, discussed and criticised, and because the students have participated in developing the mathematical analysis, they are able to follow the essential part of a new argument.

An important aspect of the application of mathematics in the project has to do with the idea of exemplarity. Mathematics has been used to identify a phenomenon which can be recognised as part of a problem concerning the production and consumption of food. The input–output figures are described for a particular farm but because of the way the problem has been addressed, the next step to a discussion of the problem in a general setting is not a large one. It makes sense to ask whether the result reflects a general tendency in Danish farming and whether the tendency is global. The students are not far from understanding the idea that the production of animal food seen from the perspective of energy, is a very expensive process. The world's problems of food production may look different as a result of participating in that project.

One possible function of mathematical modelling is to express the exemplarity of a particular phenomenon. This connects the notion of exemplarity not only to 'social imagery', which refers to the ability to imagine a social and political situation to be different, but also to 'critique'. 'Critique' was originally defined as a reaction to a critical situation, and this conceptual relationship finds an educational interpretation in the project "Energy". In fact, one result of the project was an attempt to change the use of energy both in the homes of the students and at the school. (The fact that social imagery is not the only necessary catalyst for implementing a change was also illustrated by "Energy", as the 'saving-week' programme at the school did not have any effect.) Nevertheless, I see "Energy" as illustrating the fact that empowerment via social imagination can also become an aspect of mathematics education.[16]

Originally, reflective knowing was defined in relation to the mathematical modelling process. When technological actions relate mathematical competence to a specific aim, reflective knowing has the technological way of handling the problem solving process as its very object. This aspect of reflection does not have any obvious position in "Energy", but as already indicated, I have suggested that the concept of critical mathematics education also comes to encompass the qualities of "Energy". Nevertheless, let me try to recapitulate the different possibilities. First, we could continue to maintain that an essential aspect of mathematics consists of its modelling capacity and of its potential to express something essential or perhaps unexpected. This could still be conceived of as being so important that any critical reaction has to take these possibilities into account. Therefore, critique, as far as mathematics education is concerned, has to be defined as a critique of the applications of formal methods. Second, it may be possible that the critical dimension of mathematics education has to do with the nature of the external problems themselves, i.e. problems which have been modelled. This line of interpretation – in some way the original way of defining critical education – has been diminished in my previous analysis of mathemacy. A reason for taking a step away from this 'critique of modelling', is that "Energy" illustrates the possibility of exemplarity. Investigating a particular farm creates an understanding of some fundamental global questions. In this case, the essential matter seems not to be the actual (mis)use of mathematics, but the problem area which has been modelled. Mathematics education becomes a way of grasping specific, well-described problems, and because of their value as exemplars, a particular insight might provide a global perspective and then mathematics education becomes critical. However, a third possibility exists. Perhaps the opportunity for the students to obtain a conception of their own position in the educational process is an essential aspect of "Energy", and also of critical education in general. This puts into focus the construction of the vantage points and the conditions necessary for the students to have a meta-conception of what they are doing. This means that the concept of reflective knowing is still useful, but now the concept has to be extended in a different direction. It has to do with a meta-conception of the educational process, and not just to do with meta-conceptions of the application of mathematics.[17]

CHAPTER 10

INTENTIONALITY

A critique cannot be defined solely by making reference to some particular features of a situation. To be critical means to draw attention to and to react to a critical situation which is seen as the *object* of critique but critique must also be addressed by somebody, being the *subject* of critique. 'Critique' is etymologically connected to 'criteria', meaning that critique also implies 'discerning'. Up to now, a primary concern of my analysis has been the object of critique, with the formatting power of mathematics as one focus. Reflections have been described as coming to grasp some of the sociological implications and ethical uncertainties of applying formal methods within technology. But a critical citizenship, including mathemacy as one constituent, also has a subjective dimension. To explore this means that the students (and the teacher) become the centre of the discussion.

A scene-setting can be a means to create a semantical landscape for discussing mathematical activities. The vantage points are hills in this landscape, as they bring into being components of a language of reflective knowing. But this language also has a different function, which in fact has been present in the previously described projects. A scene-setting not only provides a language about mathematics but also a language about what is happening in mathematics education and in a particular classroom. A scene-setting provides opportunities for the students to locate purposes for their own activities. It is therefore possible to extend the concept of reflecting. Reflections can address not only the social role of mathematics but also the actual teaching–learning situation; and from a vantage point the students may make their own learning process an object for reflection. This aspect I shall also try to interpret as part of critical mathematics education.

If we re-examine the descriptions of the projects already given, we find a common feature in the attempts to make what the teaching and learning are about understandable to the students. It is therefore *essential to make reinterpretations of all the given descriptions of projects*. However, not only to ask the reader to read backwards from this point, I shall in what follows try to discuss epistemological questions in order to elucidate why reflections about personal learning situations can be seen as essential to a critical mathematics education. We are going to address the subjective part of education by looking at the concept of epistemic development. Not only do we have to look for possibilities for making ideas and aims of the learning situation comprehensible to students; we also have to look for the possibilities for students to express and to make comprehensible to the teacher the way they see their ideas and hopes. In short, knowledge development must be interpreted as similar to an act, if it is to be a part of a critical awareness. This is a basic assumption, the meaning of which has to be expounded.

1. DISPOSITIONS, INTENTIONS AND ACTIONS

If it is to make sense to say that a person is performing an action, some conditions have to be fulfilled. We cannot say that a person is acting, and simultaneously say that he or she is being forced to do what actually is done. In fact, Pinocchio did not act until he got rid of the strings. Actions presuppose a degree of indeterminism (or freedom). We cannot attribute actions to mechanical systems behaving like clockwork, nor do we say that a tree is acting when it is growing. Actions cannot be described in mechanical or in biological terms; and if a person's behaviour can in fact be described in such a way, then that behaviour is not a part of his or her actions. It is not a personal action to breathe or let one's hair grow. This I see as the first essential condition for performing an act: indeterminism must exist, or, the acting person must be in a situation where choice is possible.[1]

The person acting must have some idea about goals and reasons for obtaining them. A person cannot be said to be acting if he or she does not have any idea about what he or she is doing, and why. This has to be interpreted in a broad way, because a person may well be acting even if the image of the goal is hazy and the reason for obtaining that goal obscure or merely implicit in a situation. This also differentiates action from 'blind' activity, like moving a hand, scratching one's nose, etc. Behaviour performed out of blind habit may still in a natural language tradition be called an act, but my hope is not to push any distinction to its limit, but simply to emphasise the following: Actions, as I use the term, presuppose a degree of indeterminism, i.e. a situation which makes choice possible, and also involve a degree of awareness, in the sense that the person's intentions must be present in what is done.[2]

'Intentionality' is an important analytical construct which narrows further the analysis of 'action'.[3] A characteristic of consciousness is that it can be directed towards a non-present object.[4] It is possible for me to think of a woman even if she is not present. I could think of a woman who no longer exists, such as my late grandmother, or even a person who never existed at all, my very rich aunt from America. Equally, instead of being oriented towards a person, my intentionality could be directed towards plans and ideas. It could be 'my hope that ...', 'my belief that ...', 'my dream that ...' or 'my desire that ...'. These statements express the relational constitution of intentionality. I could have the 'hope that ...' and then it will follow with a description of a mental image of some state of affairs. In all situations my consciousness is directed towards an object not necessarily present. The ability to be directed towards a non-present object defines intentionality. In this sense, intentionality describes a relationship between a state of mind and what we could call an intentional object. The intentional object is that which fulfils an intentional relationship.[5]

Let me now try to look at the relationship: *disposition – intention – action*. I will begin with the concept of *intention*. I could 'have the intention to ...' and then it would follow with a description of some type of action I intend to perform. Intentions are examples of intentionality directed towards action. (It should be noted that in this interpretation, 'intentionality' is meant in the broadest sense as referring to a variety of relationships, intentions being only one of them.) We cannot describe an action without describing the orientation of an individual. To ask if a person's intentions are fulfilled is equivalent to asking if he or she has performed certain actions. This description, however, is not quite precise; I can have the intention of becoming the next president of Mexico without having any possibility of fulfilling this intention. I could have unrealistic intentions. But if we do not take this extreme into consideration, we could expect actions and intentions to be simply related: the fulfilment of an intention becomes equivalent to performing a certain action.

A person has intentions, but I do not maintain that a person must always be aware of his or her own intentions.[6] We may be doing something without having a clear picture of what we are doing and why we are doing it. Perhaps we could express some of our intentions if asked, and this means that it is possible to make some implicit intentions explicit. Intentions for actions can be made explicit in reasons and goals.

Intentions are related to *actions* by which they can be fulfilled. I have the intention of doing so and so, and then I do it. In this sense, an intention could be the cause of an action. I do not use the term 'cause' in the same sense as we do when we say that a push on a billiard ball is the cause of the ball rolling. This is a mechanical concept of cause and effect. By using the term 'cause', I do not mean that an intention is the cause of an action in this mechanical way. It has to do with another concept of causality, and the way in which the different concepts of causality may be interrelated is an open question. Using this terminology, an intention may be said to cause an action. We have to say 'may cause' and not 'is the cause of' because we sometimes have to do with unfulfilled intentions. Intentions as causes do not necessarily produce effects.

The stipulated relationship between intention and action could even be seen as an interpretation of action. If no intention precedes an activity we would not call it an action; then it is just an activity performed out of habit or as part of a reflex. (I use the term 'activity' in a broad way to include whatever the human body may perform; that means all sorts of movements whether they be actions or not. Instead of 'activity' I could also use 'behaviour'.) I could have the intention to walk to the railway station to have a look at the trains. I could even decide to do so. Then I could perform the action of walking to the railway station. It could also happen that I needed some fresh air, and while I walked around the village it suddenly transpired that I was standing at the station looking at trains. I have performed an activity, but I would not call my arrival at the station an action. It could be shown even more directly that intentions make up a definition of an activity being an action. If I say that my intention is to walk to the station to look at the trains, it need not be the case that I have performed that action, even if, in fact, I went to the station and looked at trains. The possibility exists that I could be doing so for some other reason, or that the activity could be caused by things other than my intention to do so. Perhaps I happen to pass the station because I met a friend, and while being absorbed in old memories we happened to pass the station. The activity which I was then performing is, in fact, identical with the activity which would fulfil my original intention. However, if the activity is not performed because of the intention to do so, it does not combine with an intention to become an action.

To be an action, an activity must be related to an intention; but this needs further comment. The intention we have described as the cause of an action constitutes, so to speak, an event outside the action. But a behaviour needs something more to become an action. An action does not consist just of an initial intention, expressed before the actual activity takes place, and then the activity itself. An action is not just the combination of an original intention and a following 'movement of the body'. For instance, I do not perform the action of walking to the station if it happened just 'beneath' any consciousness. To perform an action I must, at least to some degree, be aware of the activity as part of an action. We could say that an action is composed of a physical activity, and a certain awareness of that activity. We could talk about 'intentions within the action'. These intentions are not the same as the intention that is the cause of an action, or the intention prior to the action. The intention in the action is what had to be added to a certain movement of my body if it was to be called an action. Wittgenstein once formulated the following question: What is left over from the action of raising my hand, when we subtract the actual movement of my hand? Intentions within the action are prolongations of the intention as the cause of an action.

Intentions do not spring to life from nothing. They are grounded in a landscape of pre-intentions or *dispositions*, and I shall divide these into both a 'background' and a 'foreground'. A background can be interpreted as that socially constructed network of relationships and meanings which belong to the history of the person. When we try to explain the intentions of an individual we often refer to his or her background. But the background is not the only source of intentions. Equally important is the foreground of the person. By this expression I refer to the possibilities which the social situation makes available for the individual to perceive as his or her possibilities. It is not open to me to have the (realistic) intention of being the next president of Mexico. It is not a part of my foreground and only if I were a madman would I produce intentions of this kind. The foreground is that set of possibilities which the social situation reveals to the individual. Dispositions are objectively rooted but they are not factual. Dispositions are mediated by the individual, therefore they also express a subjectivity. Dispositions are just 'dispositions', and this means that they are impossible to observe directly. The dispositions of a person are only revealed when the person comes to act.[7] Both the background and the foreground are interpreted and organised by the individual. This emphasises that 'time' does not structure the discussion of the sources of intentions. Intentions, being the cause of actions, need not to be related to the past as is the case with mechanical explanations. When we try to give intentional explanations, the future is just as present as the past.[8]

Nor is the relationship between dispositions and intentions that of cause and effect. It does not make sense to talk about dispositions as the cause of intentions, at least not in a simple interpretation of 'cause'. It is better to see dispositions as a source of intentions. Intentions emanate from the background as well as from the foreground of the individual. The individual produces (or raises, or creates, or decides about) his or her intentions and by doing so, reveals his or her dispositions. It is from an interpretation of dispositions that intentions may be raised by the individual and become causes for actions. The process of raising intentions is not a biological phenomenon, rather, intentions are identified through the decisions of the individual. It is, however, impossible to create all sorts of intentions from the source of existing dispositions. My pre-selected intentions set up limitations. Intentions can mutually exclude each other.[9]

What does it mean to have succeeded in performing an action? The answer becomes a summary of the previous notions. I have succeeded in performing an action, if I have brought about intentions from my pre-intentional background and foreground, i.e. from my dispositions, and these intentions become fulfilled by my performing some actions, which are in fact caused by the intentions, and which are performed in such a way that some intentions become part of the action itself. The story cannot end here however. Actions have effects, and it makes sense to try to interpret the concept of action also in terms of the person's reaction to these effects. This opens a cyclical process: dispositions become changed because of intentions and actions.

Dispositions are grounded in the social objectivity of the individual, and simultaneously produced by the individual, partly as a consequence of the actions performed by the individual. The success or failure of actions gives rise to modified dispositions. Through actions the objectivity of the dispositions becomes moulded, and in this way they become the new real source for intentions. The constructs disposition–intention–action become a conceptual circle for looking at actions. I have outlined this circle however in a way which needs a strong modification. I have talked about actions as an individual undertaking, but it might be better to see actions of a group as the principal conceptual unit. However, I shall not enter this particular discussion.

Just as an aside: The vindicated existence of intentionality as a unique human factor has implications for the nature of scientific explanations. Mechanical and biological explanations become insufficient to clarify all sorts of human behaviour, although some behaviour may be explained in this way. To dissect and to translate all sorts of scientific explanations into mechanical explanations has been the goal of logical positivism, and a main task of the behaviouristic tradition has been to accomplish this for psychological explanations, but neither have come to any successful conclusion. Intentional explanations, involving goals and reasons, constitute an independent class of explanations. To explain an action by making reference to goals or to anything that could happen in the future does not mean that we are trying to explain something actual by something in the future. To make an intentional explanation means to explain an act by the person's actual image of some future state of affairs and by the person's actual reasons and goals, whether they are implicit or explicit.

Even if it is important to distinguish reasons and goals from mechanical causes, a lot of interference takes place. Some behaviour may be explained in a mechanical or biological way by making reference to a person's desires and needs; the person may suppose that he or she is doing something on his or her own, while he or she in fact is governed by hidden desires and needs. The point is not to deny the relevance of such explanations in some situations but to maintain that if what we are concerned with is an action, the explanation cannot simply collapse into a mechanical explanation. Some behaviour of human beings needs intentional explanations.

The fact that intentional explanations make up an independent class of explanations, impossible to analyse in terms of mechanical or biological explanations, does not imply that a person may be fully aware of his or her own intentionality. In fact the problem of consciousness always exists. A consciousness cannot fully reflect its own state. The consciousness always sees itself in a distorted way, like being in a room of mirrors. A person cannot grasp his or her own motives, hopes, reasons and goals in total, and therefore it cannot be expected that the intentional explanation of a person's behaviour can be given by the person himself or herself. In fact, it may be highly dubious whether anybody is able to specify such an explanation. The consequence is that it is impossible to give any adequate intentional explanation of an action. Still, this does not mean that intentional explanations have to be done away with or abandoned. Intentional explanations are necessary – and necessarily incomplete.

2. LEARNING AS ACTION

As mentioned earlier, if we are to get a better grip on the meaning of critical education I shall suggest that learning should be interpreted as (similar to) an action.[10] We must therefore take a look at the relationship: *disposition – intention of learning – learning as action*. Dispositions in this context make up a totality similar to the totality from which intentions for actions emerge, and therefore we are concerned with a background as well as a foreground. The situation which could raise intentions for learning does not automatically belong to the background of the student having to do with his or her situation and social heritage. It has just as much to do with the student's possibilities in future life, not the objective possibilities but the possibilities as the student perceives them.

Intentions of learning may emerge out of dispositions, and the decisions of the learner therefore have a role to play when conditions for learning are produced. I do not see intentions of learning as being different from other sorts of intentions except that they can be fulfilled by learning activities, and learning becomes an action.[11] We could make several distinctions concerning intentions of learning. Some intentions may have to do with questions of content matter, but that is far from the only possibility. Besides having to do with the curriculum, intentions could have to do with the teacher: the student feels he or she has to do something in school because the teacher has said so. The decision about learning could have to do with the structure of schooling, like tests and marks. Or it could have to do with the student's position in the class, for instance with the competition between students. It could also have to do with student's conception of his or her future in vocational life, the student's view about the possibility of getting a job, etc. This means that intentions of learning are based on a complex set of dispositions and can guide the student in many different directions, not all of them being parallel with the direction the teacher might have expected.[12]

By describing the learning-histories of some upper secondary students, Lena Lindenskov has shown that their learning strategies display remarkable patterns in the sense that each student, when being asked, was able to explain and indicate why he or she reacted to different challenges in mathematics education in particular ways.[13] Students' meta-conceptions of mathematics create patterns for interpreting the different tasks presented by the teacher. The students identify their priorities and react according to these during the educational process. Lindenskov has used the expression 'student's curriculum' to designate such a pattern of reaction, and I see these patterns as empirical evidence of the idea that the students, during an educational process, are acting persons and that learning is similar to acting.[14] Their actions have to be understood in relation to their dispositions. The learning strategies reveal parts of the students' dispositional structures.[15]

I see learning as caused by the intentions of the learner. But here we find some divergence from the previous analysis of actions. Learning may happen even if it is not caused by the original intention. As stated, I do not call an activity an action, if the activity was caused by something different from the intention to do what the person did even if he or she comes to do what he or she intended to do. This dislocation or shift of causes prevents the activity from being an action, but nevertheless the person performs some activity. When a child learns to follow some customs, it may be difficult to analyse this in terms of learning as actions, and I shall call such a behavioural adjustment learning by assimilation or enculturation.[16] Many features of a culture may be assimilated by the individual without actually having any intention of learning anything particular. This shows that the discussion of learning as action has crucial limitations; however, I shall continue the analysis of learning as action knowing that it cannot constitute a complete analysis of the concept of learning.

Just as in the case of action, I find that learning has a dual content. Besides the intentions of learning being a cause of the learning activity, we can talk about intentions in learning as meaning being aware of one's own activity as a learning activity. Let us consider an example. A child may be playing with centicubes. This is an activity and also a learning activity, in the sense that the child may acquire some understanding of numbers and combinations. But it is not learning as action if the child does not pay any attention to the activity but performs it in an absent-minded way. The child has to be involved in the learning if the learning activity is to become learning as action. There must exist intentions in the learning. What about this normative formulation? What does 'must' mean? Something like: If we are to try to provide more meaning to the examples of critical mathematics education, an interpretation of learning as action illuminates the process.

Reflecting is *par excellence* an action. Critique is an action, and cannot be developed in school unless the learner takes some responsibility for the learning process. We can learn many things by command, but critical competence cannot be forced upon a student, and this is the reason that learning as action is an important aspect of critical education. Intentions in learning help to define learning as part of a critical enterprise. Routines may be assimilated in an absent-minded way, while a critical awareness cannot be developed as (blind) assimilation or enculturation. If learning means not just receiving information but also includes reflections, the learning has to be *performed* by the learner. Thus, I find the genetic epistemology of Jean Piaget to be misleading, insofar as it connects epistemic development with biological growth, which is made visible by Piaget's definition and use of the two concepts 'assimilation' and 'accommodation'. The biological interpretation locates the discussion of the teaching–learning processes in a context of scientific explanations similar to biological schemes of explanation and makes intentional explanations foreign to epistemology. By doing so, Piagetism creates a richness of metaphors used in describing and analysing what is going on in mathematics classrooms. Such situations become described according to the richness of the learning environments, and, most sympathetically, the best and most stimulating environments are preferred. Education is seen as a sort of horticulture.[17] But the student must have the opportunity to choose an orientation to be involved in a process of coming to know.

We can identify different fundamental conditions for a student to be involved in epistemic development, making space for critical awareness by considering the conditions for a person to perform an act. Epistemic development cannot be forced upon anybody. We cannot, as teachers or curriculum designers, implant goals in a student, nor can we implant good reasons. Goals have to be identified and accepted by the learner as of importance and reasons have to be accepted; if not they can never become the reasons of a person. And not only accepted, because the intentional orientation must be performed by the person himself or herself. A condition for a productive teaching–learning process is that a situation is established where students are given opportunities to investigate reasons and goals for suggested teaching–learning processes, and by doing so, to accentuate their own intentions and to incorporate some of them as part of their learning processes. Intentions in learning must be placed there by the learner himself or herself.[18] Therefore the concepts 'teaching' and 'learning' will lose some of there classical meaning. Teaching cannot be seen as a sort of delivery and learning as a sort of receiving.

3. DIFFERENT FORMS OF EPISTEMIC DEVELOPMENT

The sequence: disposition – intention of learning – learning as action does not indicate that students possess some explicitly formulated intentions which they actually want to bring to the learning situation. It does not make sense to talk about the intentions of students as something pre-existing even though the dispositions of the students are resources for intentions. Intentions of learning can unfold in many ways: they can be advanced, refined, restructured, remoulded, discharged, dismissed and scrapped. And this happens as part of the educational process.

The emergence of selected intentions for learning often takes place in a situation overburdened by demands. The structure of schooling makes a forceful structuring of the pre-intentional dispositions. In a normal classroom situation it is not common to see the emerging of intentions of learning as part of a negotiation, in which the teacher expresses possibilities and the students express themselves in order to grasp the situation better. Nevertheless, the activity of changing and adjusting intentions is a most common activity in school. The demands of the situation make it necessary for the students to restructure their intentions, but often this happens in a haphazard way. The adjustment of intentions does not take place as a shared experience, but as an individual undertaking. A multiplicity of different intentions, not necessarily having much to do with learning, may be set up.

The result is not that no learning takes place, but learning as action congeals as a forced activity, and the intentions in the students' actions become different from the intentions related to the learning. The demands of the situation influence the intentions which the students may add to the individual learning activities. The intentions in action become strategic. Because students always interpret the school situation in an on-going way, the demands of the situation become part of students' dispositions, and therefore part of the situation which the students take into consideration when they determine how to act. The intentions, which bring about a cause for learning (as a forced activity), are invoked by an interpretation of what takes place in the classroom. These interpretations are moulded by the students' notions of their backgrounds and foregrounds and they provide reasons and goals, but not necessarily for learning as the school expects to see it.

Learning as forced activity is found in 'standard' teaching, leaving no space or freedom open to the students. No goals are explicated, no reasons mentioned. The process is directed by a set of orders contained in the textbook and repeated by the teacher. This is a *directed activity*, or a prescribed activity, and if students are ordered to do something, they seem inhibited from taking any action at all. In this sense we are dealing with a distorted form of epistemic development. The directed activity may take different forms. It is possible to make the students 'forget' that they are without alternatives and without opportunities to see purpose and reason by making the learning situation charming and puzzling and hence, implicitly motivating. However, it is more normal to direct an activity by means of a textbook and a well-specified curriculum accompanied by tests and assessments, and the official life in classrooms rigidifies into dead and robotic figures. Normally, when just reading a textbook, it is impossible for students to figure out why they have to do this or that. A mathematical textbook is usually a carefully elaborated sequence of orders and commands, reflecting the commands put forward in the curriculum and made audible by the teacher. The classroom becomes a drill ground. The textbook may contain an attempt to motivate by giving fascinating examples and exercises, but the students are not invited to take any decisions concerning the direction of the learning process. The students' intentions are not taken into account.

The students may be guided by the teacher, the textbook and the curriculum in the same way as a group of (fatigued) tourists is guided around a town in a sightseeing bus listening to the kind voice of the guide talking about the beautiful buildings, the old churches, and about whatever else passes by. The tourists will not be involved in any action during the trip – except deciding whether to listen or not. It would, of course, be wrong to suppose that nothing will be learned, as something may be learned in any distorted epistemic development. This is similar to what is going on in directed mathematics education but with important differences. The students have no choice about whether to enter the classroom with the fixed curriculum or not. While the tourists in no way will be punished if they are not able to repeat the words of the guide, the situation is different in school. The trip may also be useful for the tourists after leaving the bus when trying to get back to some interesting places, but normally students do not have the possibility of leaving the mathematics classroom to take a more careful look at what seems to be of interest.

The antithesis of directed activity is found in *blind activity*. This is like free enterprise and not common in school, but also in this case fundamental conditions for negotiating are eliminated. It is characterised by the absence of formulated and negotiated goals as well as of explicated reasons. I mention this type of degenerative situation because it characterises some projects which have been important in developing suggestions for critical educational practice. For instance, the fundamental question raised by the Glocksee project concerns the degree to which a laissez-faire pedagogy has importance for a critical pedagogy.[19] The laissez-faire pedagogy, also incorporated in the student-centred pedagogy, has been heavily criticised from the standpoint of traditional pedagogy while critical education, in its general interpretation, has been more reluctant to refuse free enterprise in schools. This is for two reasons: First, free enterprise strikes a blow at traditional rituals in schools, and by wiping out normal standards for organising the teaching–learning process, an openness seems created. Second, free enterprise can create possibilities for students to become involved in subjects of personal importance, and perhaps with an unexpressed political dimension. However, my interpretation of epistemic development indicates limitations in this approach. To let students move around in school, accepting or not accepting different offers from the teachers (which was seen by many as a distinct feature of the Glocksee project) depending on the subject's and the teacher's ability to catch the students' attention, means a repetition of the biological interpretation of the learning process. Being attracted to a specific subject is different from seeing and understanding goals and reasons of or for that subject.

In the same way as two conditions for action are set up having to do with a degree of indeterminism and a degree of awareness, we find two pure types of distorted learning processes. In the first situation, no alternatives are left open, only a single one-way road is passable. This inflexible situation easily implies that goals and reasons are undeclared. In the second, students are put into an open situation, but this free space is unstructured and confused so that it becomes difficult to negotiate reasons and goals, and the students become left to their own devices. Neither of these extremes brings intentions into learning in a form that I see as important to critical education.

Students will enter school with ideas, hopes and expectations. Intentions are inherent within every human being. They do not have to be invented or transplanted. What is significant is the continuous, almost strange, development of a new richness of intentions. This does not mean that intentions cannot be moved, changed, developed, made obvious, etc. This is what has to be done in education. But the demands of the situation in school too often result in *broken* or *ignored intentions*. When students' intentions are ignored, it seems impossible for students to perform actions which could fulfil negotiated intentions. This could happen by not giving the students any possibilities to express goals and reasons, or it could happen by not offering any reasons for what is going on. In short, it could happen as the result of directed activity. Broken intentions are not the same as *modified intentions* or *integrated intentions* or *shared intentions*. The teacher has ideas and plans, the students as well, and no parallelism can be presupposed. To share intentions presupposes a complicated procedure involving the imagining of different goals and reasons. The relevance of dialogue and negotiation in critical education has precisely to do with that modification. By the activity of sharing intentions, the students may come to act as a group and add a new force to the dynamics of learning.

We have to show respect for other people's intentions, but we need not assign them any sacred position. Intentions are determined by a person's conception of goals and reasons, which may be foggy and may not be consistent or well-elaborated. Intentions are questionable but not ignorable, and the teacher has to challenge and to try to understand the different intentions of the students – as well as the students having to question the goals built into the curriculum and the teacher's intention by doing this and that. This process can create collaboration in the classroom, which forms the basis of the modifying of intentions. Sharing intentions involves the setting up of alternative goals and criticism of reasons for doing so. Reasons and goals have to be put into words or made visible in terms within the horizon of the students. (As described here, the sharing and modifying of intentions seem to be a rational process, but in classroom practice they may take a variety of more complicated forms.)

In traditional mathematics education intentions are seldom shared. Negotiations are cut off by: 'Today we have to learn about...'. Possible intentions behind the series of educational commands are not made comprehensible. The student is left somewhat as a soldier in a trench at the front, without knowing about his actual position or about the next movement of the army. The soldier has no possibility of getting an overall image of the strategic situation. If he starts to carry on his own warfare in the light of his interpretation of the situation, it will probably not make any sense with respect to the overall strategic situation. The soldier cannot act, only follow commands. However, the soldier may develop his own interpretation of his personal situation. He may develop his own intentions, like keeping himself in the best possible physical condition, trying to get a better ration of food, improving his ability to play chess, etc.

This exemplifies a common phenomenon in classrooms: the development of *underground intentions*. Those intentions are not shared in the 'classroom public'. When the intentions of the students are ignored, the consequence is not that the students become emptied of all sorts of intentions and prepared to accept the delineated aims and reasons. Instead, a richness of underground intentions will emerge and the student's curriculum will easily rise to confront the official curriculum. The students cannot act as part of a common learning process, they become stripped of power when they enter the classroom, but they create a new space for possible actions.[20] Examples of underground intentions can be: How to hide the fact that homework has not been done? How to get top marks? How to make more friends? How to get into the same working group as Jane? The student can invent a great variety of underground intentions giving meaning to life in school when the curriculum itself has become stripped of meaning. Traditional mathematics education, therefore, is characterised, not by the modifying or the integration of intentions, and normally not by the substitution of any intentions, but by the multiplicity of unshared intentions. That is why underground intentions flourish in an epidemic manner in classrooms, like hopes and dreams among soldiers near the front, or like potato shoots in a dark cellar.

Some of the unshared intentions may be characterised as *instrumental*.[21] They fit very well with the demands of the school structure. To behave in accordance with the expectations of teachers and school may well be set up as a goal of the students, meaning that some excellent performances in a subject may not be determined by an interest in the subject itself, but are a consequence of a wish to be well-behaved pupils. In fact, students may realise that the best possible strategic action in school is to get the best possible marks, whether the subject matter is of any interest or not. The important thing is to obtain the best possible position when the rush begins towards further education. This means that instrumentalism becomes a part of the students' own decisions. In this sense, unshared intentions can be the cause of the sort of learning we have schematised as a forced activity which may sound contradictory, but students can internalise an external restructuring of intentions, and by so doing carry out instrumental actions.[22]

4. PERSONAL FATALISM, SERVILITY AND ACHIEVEMENT

Let us take a closer look at the phenomenon of forced activity. In this case the students have no opportunities to pursue pre-formulated intentions, and they do not know whether they might be valuable or not. Some may develop an instrumentalism which fits the 'logic of the school', others may be lost, and the result may be *personal fatalism* or the production of negative self-esteem. It is indicated by phrases like: 'I am not able to do ... ', 'I am not interested in what they are doing today', 'I have to do so and so, but I do not know why', etc. Personal fatalism is produced by a devaluation of students' ideas and goals and by ignoring and not valuing their abilities. Personal fatalism is caused by the elimination of the possibility for the student to become involved in learning as action.

The possibility that a well-structured mathematical curriculum can be obstructive to learning has been underlined by Ubiratan D'Ambrosio as part of his concern for the ethnomathematical approach. D'Ambrosio finds that the learned mathemacy can eliminate an already existing spontaneous mathemacy. D'Ambrosio finds that even if a person is able to manage (informal) calculations and operations with geometric forms, this competence becomes easily suppressed when facing similar mathematical tasks formally dressed up. Facing this stiff formality a psychological block is created, and a barrier is raised between mathematical formalism and implicit and wordless mathematical competence.[23] In ethnomathematics it has been pointed out that personal fatalism may be produced by the institutional ignorance of an already existing competence integrated into the pre-understanding of the student and thus belonging to the student's culture. This pre-understanding will probably not be fully verbalised, and certainly not verbalised in the language of school mathematics.

Personal fatalism is one possible characteristic of what I have called the dispositions of a person. Dispositions are determined by the possibilities which the social situation provides for the individual, and by the way the person interprets those possibilities. An experience in school may support feedback from the teaching–learning situation to the dispositions of the students and cultivate personal fatalism. Personal fatalism can take a variety of forms, and one of them I see as a *servility* towards technological questions and towards those who can manage such questions.

This possibility has to do with the assumed formatting power of mathematics *education*: servility of students comes into being when pre-understandings are ignored. The formatting power of mathematics education can be discussed in terms of broken intentions, modified intentions, underground intentions, instrumentalism and personal fatalism. It can be discussed as a restructuring of the learning dispositions of the students, that is, a restructuring of the sources from which students bring intentions into their learning processes. However, the formatting power of mathematics education is not a feature which has to be prevented – this is an impossible task. The theoretical aim must be to clarify this formatting, and the task of a critical mathematics education must be to see this formatting in the light of concepts such as critical competence.

I find that students are able to learn almost anything, if they have reason to do so. This is a consequence of seeing learning as being (partly) determined by a decision of the learner.[24] But if learning has to do with action and is specified by intentions, it becomes important to interpret the performance in school in terms of the students' views of their future. Formal mathematics may seem hostile and eccentric to many students, because it is difficult for them to locate themselves in a (future) situation in life, which they hope to get into, and in which formal mathematics met by them in school plays a vital role. Therefore, it is impossible for the student to identify any personal reasons, except instrumental ones, for learning abstract calculations. When we discuss a student's performance in school, we do not (only) have to look at the student's background, but at his or her foreground as well. The future gives reasons and dreams, or destroys them. Actions, and therefore epistemic development, are also oriented towards the person's actual perception of the future.

Therefore, if the student's opportunities for learning can be eliminated in the same way as opportunities for actions, the concept of achievement needs reinterpretation. The analysis of achievement makes a global sociological perspective necessary. The possibility for a student to maintain a goal in school will partly be based on the set of future (work) possibilities which the student is able to perceive as his or her own. If the student knows that he or she probably will become a worker, because that is what everybody in the family has been, it is difficult to attach personal aims to the learning of abstract mathematics. And in fact, in different societies very different sets of possibilities are made visible for students depending on their social background. The consequence of this is that a difference in achievements cannot only be analysed in terms of the (lack of) richness and stimulation which the social background may or may not provide; equally important is what society allows the students, from a certain social background, to conceptualise as their future. My point is that when students from a certain social group are identified as achieving less in mathematics than others, the explanation for this phenomenon can also be found in the difference in the future of the different social groups of students. This future is present in the dispositions of the students. If, for instance, different attitudes of girls and boys towards a certain subject are identified, the explanation of this phenomenon has also to be expressed in terms of their perceived futures in society. Similarly, if a society realise racism in terms of differences in achievement in school depending on colour of skin, this phenomenon also has to be seen in terms of the future which the society provides for the different groups. It all has to do with the fact that students interpret life in school in the light of their vision of personal opportunities in the future. In this sense, achievements in school reflect the dispositions of the students.

Complete rationality presupposes that the person is transparent to himself or herself – but this cannot be assumed. An essential point, however, is that actions which might look irrational from one perspective can have a coherent meaning from a different perspective. It might seem irrational that students sometimes ignore a subject. They do not do their homework, put their head on the desk when coming to school, and do not even try to conceal their lack of interest. Nevertheless, this 'irrational' activity can have a rational content. To ignore a subject might be a way of protecting oneself from demands from a teacher, from a schoolbook and from a curriculum which seem overwhelming. To prevent complete defeat, i.e. not to be able to cope with problems they want to be able to handle, the students decide not to pay any attention to the subject at all. This becomes a last defence towards personal fatalism.

5. SOME COMMENTS ON THE PROJECTS

The interpretation of learning as action makes it important to return to the projects: "Energy", "Our Community", "Family Support in a Micro-Society", "Constructions", "Golfparken" and "Economic Relationships in the World of a Child". Initially, reflections were described as having to do with technological actions, and, because we are concentrating on mathematics, the object of reflections becomes applications of mathematics. Reflective knowing has been recognised as an essential element in mathemacy, integrating mathematical, technological and reflective competencies. As previously explained, mathemacy designates a complexity similar to that of literacy, and, like literacy, mathemacy may provide educational sense to a notion as 'critical citizenship'. This is the reason that mathemacy might include a beginning for critical mathematics education. This line of analysis has, in turn, pointed towards an *object* of critique.

The perspective of this Chapter, however, has been the *subject* of a critique, and this adds a new dimension to the concept of reflections. Reflections can be oriented towards the situation of the subjects themselves, and in the case of education, this means the situation of students and teacher: Why do we have to learn this in this specific way? It is my assumption that the subject of critique has to be added to our conception of reflections in order to provide a coherent meaning for critical mathematics education. The subjective aspect might be as essential as the objective one: It is not only important that mathematics education provides the opportunity to understand essential features of society and the role of mathematics, but also that different intentions of the learning are negotiated and shared, and that the students in this way become actors of the learning process – and not blind passengers.

This brings us back to the concept of *Mündigkeit*, injected into the educational discussion by reference to Adorno. *Mündigkeit* contains a double orientation which reveals the subjective as well as the objective orientation of critique and, consequently, the dual nature of reflections: To obtain *Mündigkeit* you have to be able to take well-grounded judgements, and you have to do it by 'authority'. This duality is the strength of this notion which originally characterised critical education in its effort to conceptualise an 'education after Auschwitz'. I find it important to keep this duality as part of critical mathematics education. But this does not mean that I want to discard the characteristics of reflective knowing and of mathemacy as outlined in Chapter 6, but these formulations have to be rethought in terms of reflective duality.

I believe, however, that both the objective and the subjective dimension of critique have already been illustrated by the projects discussed. The projects, therefore, reveal a lack in the original analytical approach which I have now tried to remedy by paying attention to 'intention'. This makes it necessary not to invent new projects, but to take a fresh look at the 'old' examples, although, in this instance, I shall restrict myself to making a few comments to illustrate what could be meant by expressions like 'shared intentions', 'intentions of learning' and 'students as (responsible) actors in the educational process'. Additional meaning is also provided to the concepts of 'setting a scene' and of 'vantage point', previously described as 'semantical hills' of a language of reflections. To negotiate intentions makes it necessary to talk about the possible directions of an educational process. This presupposes a terminology in which to express possible intentions-of-learning and possible intentions-in-learning. This is precisely what scene-setting and vantage points might provide: ways of explaining and negotiating the meaning of an educational process. A scene-setting is artificial, but this artificiality can become a rational way of expressing the meaning of an educational activity.

In "Energy" the scenes were carefully set up. First, it concentrated upon the input–output figures for energy as part of the students' own activity of cycling. Then an investigation of input–output figures of farming took place. Those two scene-settings were not just some motivational device, but also a way of making it possible for the students to get an idea of input–output figures and to let them understand why *they* have to study this phenomenon. Naturally, it would be possible to inform the students about the problems of the use of energy in farming, but this does not bring the students and the teacher into a process of sharing intentions. A scene-setting is one strategy to create a space for epistemic actions: – Why do we have to weigh out breakfast? Is it OK if I do not eat breakfast food before coming to school, but just have a bag of chips? To answer such questions presupposes a reference to ideas of what the whole project is about. And in order to make sense, these explanations cannot be given in a language referring to phenomena beyond the horizon of the students.

"Our Community" provides a straightforward educational interpretation of 'students being involved in the learning process'. An initial task of the students was to apply for a job at the Hinnerup District. It was obvious that it was a 'pretend reality', even if the students had to look up the jobs they wanted to apply for, to write their applications, to go to the job interview, and to wait for the result. The artificiality of the scene-setting, however, did not make things less real. In this project the actual foreground of the students was established as part of the intentional content of the education. To decide about which jobs to apply for, the students became provoked to take a realistic view of their own future. Hopes and dreams about work possibilities became mixed up with reality, and from this foreground they could bring their intentions into the learning process. Also "Family Support in a Micro-Society" involved the students, in this case as members of the government of different districts having to decide upon, and to carry out, a distribution of support to the families living in the districts. The students were assigned a specific role, and to understand the significance of this role became a way of understanding the personal significance of the educational tasks.

The essence of "Constructions" does not seem to have much to do with coming to reflect upon the use of mathematics, although I have tried to indicate the possibility by showing how mathematical archaeology could have continued that project. The children could have come to see what sort of mathematics, in fact, might go on behind their (technological) activities: How to construct a bridge as stable as possible? Why think in terms of triangles? An important experience for the teachers involved in this project was the extraordinary energy which the children put into their work, even though their possible intentions were not explicitly negotiated. This indicates an important idea: Intentions can be shared even if they are not expressed in words. It is possible to *show* intentions and to *see* intentions without the intentions being actually described. Up to now I may have indicated the process of sharing intentions as a (semi)rational process which presupposes that intentions are described. But this notion of negotiation is not necessarily the only relevant one. Sharing intentions can take place in different ways. This is also emphasised if we look at "Golfparken", which was embedded in a careful scene-setting: How to improve the landscape outside the school. (Nevertheless, in this case it was more difficult for the children to 'get into' the project.) So, when I talk about the negotiation of intentions, I do not refer to some necessarily explicit discussion, but to the possibility of seeing the meaning of the teaching–learning process and for the students to perceive their positions in this process.

Teaching has to create a richness in which it is possible to find and to share goals and directions. The intentions of the suggested activities must be accessible to the students, and scene-setting is an attempt to create possibilities for negotiation (to be understood in an extended sense) in a language accessible to the students. However, negotiation presupposes a sort of parallelism. And not much has been said about the possibility of the teacher coming to understand the intentions of the students. How is it possible to involve students in the process of scene-setting? Is it possible for the students to invite the teacher into their scene-settings? These questions can also be discussed in relation to the projects already described as, for instance, "Economic Relationships in the World of a Child".

An appropriate direction of the analysis is to investigate the examples further in order to develop the subjective aspect of critique in a balanced conception of critical mathematics education, incorporating the duality of reflection. But I shall not make this return journey to educational practice. Instead, I shall make an excursion into epistemology and take an additional look at the concept of knowing.

CHAPTER 11

KNOWING, AN EPILOGUE

In the previous analysis I have used the concept of knowledge several times, and I have outlined different types of knowing related to mathematics education: mathematical, technological and reflective. But I have not tried explicitly to define 'knowing'. I shall not remedy this lack of analytical consistence in this final chapter, although I shall investigate the concept a bit further, and by doing so try to relate better the concepts of 'reflections', 'critique' and 'knowing'. My aim is not to add to an interpretation of critical mathematics education, but to take a closer look at a part of the developed conceptual framework.

Although 'knowing' is a problematic term, 'reflective knowing' has been one of the key-terms in the previous analysis. We could think of 'knowing' as a controlled concept, but any attempt to outline an exhaustive interpretation of the term reveals its open texture. Delving into the concept, we come to a wide variety of notions and aspects – also aspects quite different from those normally included in an epistemology. In this sense 'knowing' reveals an explosive nature. I want to step away from my analysis of critical mathematics education by elaborating upon this aspect of the conceptual underpinning of my analysis – and 'knowing' is only one of the explosive concepts upon which my analysis is built. I do not feel any safer staying in a house built of bricks each of which can explode. Anyway, let me begin in a seemingly controlled corner of this building.

1. KNOWLEDGE – A CONTROLLED CONCEPT

The assumption of the *homogeneity of knowledge* states that it is possible to integrate all sorts of knowledge into one unified system, and this thesis finds support in both rationalism and empiricism. Rationalism points at *ratio* (i.e. reason) as the uniform source of knowledge, while empiricism provides the senses with this unique role. According to rationalism, the mind will not contradict itself if it is able to purify itself by shedding all preconceived opinions. Traditions and norms give rise to bad habits of thinking: they may cause a repetition of false assumptions and an imperfect deduction. Instead of supporting any improvement of knowledge, traditions and habits produce a deterioration of knowledge. To produce knowledge, the individual has to get rid of all ideas smuggled into the mind by social life. The mind has to be properly cleansed to get a correct and fresh start. This is how René Descartes describes his scientific and philosophic method. The first problem, then, is where to obtain a starting point. Descartes tries to find firm ground in a simple proposition not needing any supporting argument. According to him, when you understand the proposition, you also understand that it must be true; and accompanying this understanding you have a criterion of knowledge. Then by starting from simple and basic truths deduction can proceed in small but distinguishable logical steps towards complex statements of truths. No truth is so remote that it is impossible to reach by strict deduction. By this formulation, Descartes claims that a rationalist system of deduction is complete, and *ratio* becomes omnipotent. All truths deduced will fit into the collections of already established truths. The system of knowledge is consistent.

Also, empiricism conceives the thesis of homogeneity of knowledge to be true. If we simply perceive, we shall be able to accumulate sense-data, and empirical facts cannot contradict each other.[1] One distinct proponent of the empiricist interpretation of the thesis of homogeneity is Carnap. He has stated that an analysis of the concepts of science will show that, no matter which subject the concepts belong to in the traditional classification of physics, biology, chemistry, psychology, sociology, etc., they will find their place in a unified system containing all scientific knowledge. Concepts in, for instance, biology will be revealed as molecular, but when decomposed into atomic parts their empirical significance will be obvious. The truth value of an arbitrary proposition will then be determined by the truth values of its atomic parts.

The thesis of homogeneity of knowledge has a long history in philosophy. It was anticipated by Plato in the idea that the objects of knowledge are ideas, not belonging to our actual world. If a person obtains knowledge, it must be about that eternal world and his or her knowledge cannot develop into any kind of contradiction. Plato's thesis of homogeneity is expounded in his discussion of knowledge, which in fact began a whole tradition in epistemology by raising the question: What has to be added to 'true belief' in order to constitute 'knowledge'?[2] An answer to this question is synthesised in *the classical definition of knowledge*, according to which three conditions must be fulfilled if a person is said to have knowledge. The person A knows p if and only if: A believes that p is true, A has sufficient reason to justify p, and p is true. In short, knowledge is justified true belief. As I see it, this classical conception of knowledge is not 'implied' by the thesis of homogeneity, nevertheless it reveals such a conception because knowledge becomes linked to Truth and true statements cannot be expected to contradict each other.

The notion of knowledge as justified true belief has developed hand-in-hand with foundationalism in epistemology, which maintains that a firm ground for knowledge can be found: according to rationalism, the foundation is to be found in simple axioms, the truth of which can be grasped by an intuitive act, and according to empiricism we find the foundation in simple statements about sense experiences. This brings into being an assumption of the existence of an *authorised body of knowledge* (a body of 'truths'). This assumption is a different way of stating the thesis of the homogeneity of knowledge. The possibility of different competing bodies of knowledge is ruled out, and 'knowledge' comes under control.

It is interesting to look at the concept of 'critique' which is entailed in the classical definition of knowledge. Kant has emphasised the necessity of a critique of knowledge in terms of a critique of pure reason. Such a critique of knowledge can be characterised as a once-and-for-all critique. Once it has been carried out we see the sense in which it is possible to obtain knowledge, and the development of knowledge becomes a cumulative process. The idea of a once-and-for-all critique is part of the classical definition of knowledge, appealing to a search for an authorised body of knowledge. The essence of critique then becomes to clarify the ground upon which to erect the authorised body of knowledge. Critique becomes an *a priori* activity and separated from the accumulation of knowledge (regardless of whether the accumulation is described in rationalist or empiricist terms). As a consequence, the basic 'logic of education' becomes that of delivery of knowledge.

The idea of the possibility of a once-and-for-all critique establishes critique as a philosophical practice separate from education. Critique enters philosophy as the task of identifying a foundation of knowledge, and the individual epistemic subject becomes devoid of a responsibility for critique. Critique does not move into education, nor does it become a part of the concept of learning. In the whole tradition of education we find a strong impact of the assumption of the existence of an authorised body of knowledge.[3] If a knowledge authority exists, the content of education becomes a repeat of some of the most simple and basic parts of the body of knowledge, and by means of teaching, these parts are brought into the classroom. If critical education is to make sense, we have to see critique not exclusively as an *a priori* but also as an *a posteriori* necessity.

2. KNOWING – AN OPEN CONCEPT

To give up the idea of the existence of an authorised body of knowledge means to look for knowledge without foundations. Therefore, absolutism becomes substituted by fallibilism, and the idea of situating knowledge near the concept of truth must be questioned.[4] 'Knowledge' becomes an open concept, and so does 'coming to know'. So, what could be interpreted as 'knowing' if we do not presuppose the existence of an authorised body of truths? Does any interpretation make it possible to use the word 'knowing' without a claim of universality? According to the classical definition, a statement about knowledge is a sort of description. The proposition 'A knows p' comprises a description of a specific belief (A's belief in p), a description of a structure of rational support, i.e. a description of some logical relationships (A's sufficient reasons), and it comprises a description of some metaphysical relationships between a linguistic structure and some states of affairs (the truth of p).[5] However, if a statement about knowing is not a description, what else could it be then?

Following ideas about the existence of a variety of language games and the multiplicity of functions of such games, I shall try to combine knowing with a 'willingness to argue in favour of a specific position', i.e. I draw an analogy between 'knowing' and 'promise'. If a person A (let A be a woman) promises p, it is not only a report about some state of affairs. A gives a description of the content of the promise but is not just describing this content. More is involved than simply uttering a sentence. A is *promising*, and by so doing she performs an act by means of her use of language. That a person knows a proposition p can be interpreted along the same lines.[6]

If A maintains that she knows p, she is also offering assurance that she is in a position of a sufficient background to maintain p. She must be sure that other persons can act as if p is true. But to 'be sure' is not a metaphysical concept. It cannot be defined by reference to some chain of rationalist steps of thought or to a set of sense impressions. To 'be sure' is not a measure of a degree of certainty, which a person is able to obtain and subsequently to make a report about. To 'be sure' means that the person finds she is able to conceive and imagine all situations relevant to make the knowledge claim, and that she is willing to argue that she has been able to conceive all relevant possibilities. To 'be sure' is not only a psychological description, it is a part of a promise. 'Knowing' therefore has also an ethical dimension. That A knows p, does not only mean that she is put under an obligation to discuss reasons, because knowing involves to 'be sure', she is also put under an obligation to show what she in fact is promising when she declares that she knows p. She must be able to give an exposition of her interpretation of 'the truth of p'.

The classical definition of knowledge makes references to a mental state of affairs, to logical structures, and to some metaphysical relationships, while this definition refers to a social context. From a solipsistic standpoint it does not make sense to promise anything. A promise presupposes the existence of interpersonal relationships (not taking into consideration the possibility for making oneself a promise). The same is the case with regard to knowledge. To express knowledge presupposes a social context.[7] This is a principal insight to be gained from comparing 'knowledge' with 'promise'. The expression 'A knows p' is neither less nor more specific than 'A promises p'. 'Knowing' has an open texture. We cannot define 'knowing' by reducing the notion to hard-boiled concepts, we can only explain 'knowing' by relating that open concept to other concepts just as open, hoping that such a network of semantical similarities may bring about a clarification.[8]

As a consequence we have to face the possibility of *knowledge conflicts* – which means situations where contradicting knowledge claims are made and no procedure for solving the conflict is available. Knowledge conflicts cannot be ruled out, as claimed by the classical interpretation, nor can they be labelled as unimportant and trivial, as claimed by a relativist position. A knowledge conflict belongs to the very heart of knowledge development. The possibility of knowledge conflict is a consequence of interpreting knowing as a sort of promise, made by specific persons in specific situations enclosed by a specific horizon. By 'knowledge conflict' I do not mean a belief conflict emerging when two persons have different opinions. That is what the relativistic interpretation degenerates into. A belief conflict is not problematic from an epistemological point of view – people may well be right in adopting different beliefs. But a knowledge conflict leads into a sensitive epistemological situation. It cannot be ignored. It has to be settled. However, a knowledge conflict cannot be solved by any simple empirical test. It has to be solved in terms of negotiation.

The existence of knowledge conflicts means a breaking down of the assumption of the existence of an authorised body of knowledge. An implication of this is that the possibility of a once-and-for-all critique cannot be maintained. A critical investigation of a knowledge claim becomes a permanent necessity. There is no knowledge without preconception and prejudice. We have no foundation, no homogeneity and no authorised body of knowledge. We have to introduce a perpetual critique.

To be inside preconceptions and prejudgements is an ongoing human condition. This is a consequence of being forced to live *in* language and (therefore) inside a horizon of pre-understandings, the limits of which are impossible to conceptualise. To imagine the possibility of a purifying act in terms of a sweeping critique is impossible, unless we return to a conception of knowledge as being controlled. So the unalterable condition is to live within a horizon, but this includes an obligation: Try to shed prejudgements.[9] This indicates the importance of critique as an enterprise which always has to be carried out along with the development of knowledge. Critique is an ongoing demand. This is a consequence of giving up the assumption of the existence of an authorised body of knowledge and accepting 'knowing within a horizon' as a basic epistemic condition. We always have to make a knowledge claim an object for further investigation. We have to negotiate what is conceived as the content of the knowledge claim and what is the basis for 'being sure'.

To move from the idea of the existence of a knowledge authority to the possibility of knowledge conflicts, brings critique into education. The thesis of homogeneity, in one variant or an other, has dominated epistemology for about 2,000 years. The first real attack on this thesis was launced in the 1940's by the suggestion that knowledge claims may be similar to promises. At the same time critique became considered within education. The idea that we cannot set up any once-and-for-all critique but that critique must be described as an ongoing necessity, makes it important to develop learning theories which view the learner as an agent in the learning process. This establishes both 'intention' and 'reflection' as educational concepts. We cannot force anybody to become a critical subject. Critique and forced activity are foreign to each other. It is impossible to lure anybody into being critical. To take up a critical attitude presupposes a decision made by the learner. The union of 'knowing' and 'critique' as part of an educational epistemology is made possible by the introduction of 'intention' and 'reflection'. What I have called reflective knowing designates an outgrowth of this union, but we also have to remind ourselves that the term 'negotiation' has to be used to characterise this union.

3. MONO-LOGICAL EPISTEMIC THEORIES

The prefix 'mono' indicates that the epistemology concentrates on the individual and maintains that the individual possesses adequate resources for coming-to-know, while the prefix 'dia' (in dia-logical epistemic theories to be discussed in the next section) indicates that the basis for coming-to-know is found in interactions. In this section I concentrate on *monologism*.

According to the previously mentioned rationalist tradition, knowledge is acquired by an individual act. In developing new knowledge, the only important competence to use is *ratio*. When deduction has to find its beginning in basic truths, some truths have to be perceived by an intuitive act; and in the rationalist tradition, intuition is seen as an individual act. The rationalist enterprise then takes the shape of the growth of knowledge by logical deduction from this *origo*, and deduction also is an individual undertaking. When *ratio*, including intuition and deduction, is conceived of in this way, interpersonal relationships have no essential role to play in the development of knowledge. Knowledge development and communication become separated. Communication is not integrated in the definition of knowledge and how it progresses. In this sense, I characterise a rationalist epistemology as monological.

Monologism is also a feature of empiricism, stressing the source of knowledge to be the senses (and nothing but the senses). Sensations are fundamentally individual. I do not share my sensations with anybody. Nobody can feel my pain. Sensations are individual, and consequently so is the empiricist foundation of knowledge. In the empiricist interpretation of knowledge development, a 'mechanicism of association' has been suggested. We get a mixture of sensations but, because we are able to see similarities, we make associations, and from these associations objects are constituted and regularities are identified. However, knowledge still rests upon the associations we find between our private sensations, and to associate seems also to be an individual task. This means that monologism also characterises the empiricist interpretation of knowledge development.

In being monological, rationalism and empiricism are similar, but in relation to the nature of the activity of an individual they are different. For an empiricist, we use our senses to obtain knowledge, and the act of perceiving is a passive one. I do not have to 'do' anything to look out of my eyes. If, for instance, I perform an act by making some interpretation, it may disturb the organisation of my sensations. I may be tempted to inject some of my prejudices and to 'see' according to some of my well established (bad) habits of association. It is essential in making an impressionist painting to neglect ideas about how things are supposed to be and to try to relate the activity of painting solely to the actual impressions. The difficulty of doing this is, according to Hume, for instance, caused by the strong force by which habits permeate our perceptions. If we are able to eliminate such 'ideological' influences, we shall be able to grasp pure facts in the shape of sense-data. They are not mediated by any socially constructed system such as language, for instance. Contrary to the empiricist activity of passive perception, the activity of deduction, as developed by rationalism, is an act performed by the individual and this means that rationalism has incorporated a sort of constructivism. The epistemic subject of rationalism is active.

Monologism is also found in a modern theory of knowledge. It is Jean Piaget's intention in establishing a genetic epistemology, to draw upon both the rationalist and the empiricist traditions, without degenerating into entirely either one or the other. Piaget makes a distinction between two kinds of experience: one having to do with the physical properties of objects, the other with the operations which a subject may carry out on objects. This distinction is fundamental to Piaget's identification of the genetic roots of pure mathematics. Piaget maintains that mathematical knowledge is not created by an abstraction from physical properties of objects. This is opposite to the normal empiricist interpretation of, for instance, the origin of geometric concepts: the concept 'plane' is an abstraction from physical planes, and numbers must be 'numbers of something'. Piaget, however, agrees with empiricism with respect to mathematics having an empirical basis, but instead of focusing on the subject's observation of physical properties of objects, Piaget concentrates on the subject's experience of what it is possible to *do* with objects. His thesis is that mathematical knowledge is founded in actions, i.e. the operations, which are carried out on objects.[10] Piaget agrees with rationalism that the development of mathematical knowledge is produced by a rational act, but does not find this to have much in common with logical deduction from intuitively perceived simple facts. Nor does Piaget find it to be an inductive act. Instead it takes the form of reflections on the operations (on objects). By making these operations the object for reflective abstractions, mathematical knowledge will grow. Immanent schemes of operations become explicit logico-mathematical patterns of thought.

One problem related to Piaget's thesis is that operations on objects seem to be individual and altogether different. How can such operations become a foundation of mathematical knowledge, structured in non-personalised and general terms? However, Piaget locates something quite common in operations with objects, and these common features constitute the foundation of mathematical knowledge. Piaget separates the 'psychological subject' from the 'epistemic subject' and maintains that the general nature of mathematical knowledge is rooted in the epistemic subject stripped of all individual features.[11] When asserting the common nature of reflective abstractions, by locating them in the epistemic subject, Piaget follows rationalism, which asserts the common nature of *ratio*, but with the difference that the Piagetian *ratio* acts upon something different from the rationalistic *ratio*.[12]

Classical empiricism develops a 'copy-theory'. The subject is passive, it receives impressions, and the impressions become condensed into concepts. Piaget combines rationalism and empiricism by locating an empirical source of mathematical knowledge and by describing rational means of developing mathematical knowledge; and from this combination Piaget develops his constructivism. An activity on the part of the subject is presupposed in order to obtain something to reflect upon, and reflective abstractions not only preserve, but also refine, the active part of the rationalistic *ratio*.[12] Reflective abstractions are the vehicles for the development of mathematical knowledge.[13]

Piaget's genetic epistemology sustain a monological interpretation of knowledge development. The operations of the subject are not defined in relation to other subjects but as individual manipulations on physical objects. Also reflective abstractions, performed by the epistemic subject, are isolated. No communication with any other epistemic or psychological subject is needed for such abstractions to be carried out. Also this assumption is underlined by the Piagetian idea that the sources of mathematics and logic are 'deeper' and different from the sources of language. There are similarities between the *ratio*, as described by rationalist philosophy, and the epistemic subject. Both are of a common nature and both, in their purest form, work in isolation. Reflective abstraction provides the cornerstone for genetic epistemology as deduction does for rationalism and induction for empiricism. And the process of reflective abstraction repeats the monological nature of deduction and induction. Piaget's constructivism becomes monological.

A main perspective on mathematics education has been provided by Piaget's monologism. In primary school teaching monologism puts the intellectual development of the child as the main focal point. The child must be placed in the most stimulating environment, and, with reference to genetic epistemology, it is possible to specify important features of this environment. It should be possible for the child to be involved in concrete operations, facilitating the development of logico-mathematical patterns of thought. The progressive and experimental mathematical classroom is inhabited by children drawing, cutting, pasting and building. This has created busy and lively activity. And certainly it has created discussion and interaction between children, and between children and the teacher. And yet this activity has been interpreted by a monological epistemology, meaning that communication is seen as having to do only with the atmosphere around the learning process. Communication is reduced to a pedagogical and methodological implement.

If a proper source of mathematical knowledge is identified in these terms, critique becomes only a prerequisite for students to reconceptualise their existing constructed mathematical competence, and not an activity related to the educational situation in which they are involved or to the social role of mathematics. Critique as self-critique, oriented towards the (individual) construction of mathematical knowledge, is the result of epistemological monologism. Monologism makes it difficult to see interaction in education as a precondition for a necessary reflection upon the content matter in question. Therefore, the monological perspective, which characterises most offshoots of the Piagetian interpretation of activities in the mathematical classroom, makes it difficult to re-think critical education. Monologism excludes criticism as an educational obligation oriented towards 'the politics of knowledge'.

4. DIA-LOGICAL EPISTEMIC THEORIES

As an epistemic concept, 'dialogue' belongs to the same logical type as 'perception' and *'ratio'*. It is my thesis that we have to turn to a *dia-logical* epistemology to integrate critique as an educational concept in order to achieve an epistemic understanding of critical mathematics education. This also means that we have to struggle with the use of a cliche. 'Education as dialogue' has been the slogan behind which many happy and progressive educators have lined up. But although sugar-coated, 'dialogue' contains an essential epistemic point. In this context my use of the term 'dialogue' has much in common with the term 'negotiation'. The establishing of 'dialogue' as an epistemic concept is implied by giving up the thesis of the homogeneity of knowledge, and accepting that contradictory knowledge claims can rightly be made with the consequence that knowledge conflict becomes a reality. To see 'knowing' as similar to 'promise' reveals the social content of knowing. Just as 'promise' does, so also 'knowing' presupposes an interpersonal relationship, and this becomes a basic condition in any education which refuses 'banking', to use a notion coined by Freire as a description of a teacher–student relationship.

Knowledge conflict is a sensitive epistemic phenomenon, and cannot be solved by adding new information, collecting more observations or by performing more careful calculations. Knowledge conflict has to be handled in a different way. Critique and reflection are needed. From knowledge conflict, we may hope to develop new concepts and to be able to reflect upon knowledge already held. If knowledge conflict is to enter into a dynamic process, its critical and dialogical nature has to be emphasised. If A knows p, B knows q, and p and q are contradictory, then any progress will depend on the interaction between A and B.[14] This is the only way to handle a situation with two partial conceptual systems. Knowledge conflict indicates that some conceptual frameworks are incompatible, perhaps inconsistent. Therefore, knowledge conflict must result in a change of concepts. No mechanical procedure exists for doing this. The upheaval of a knowledge conflict cannot be the result of pure reasoning or of some carefully carried out experiment. The only way forward is negotiation.

If we abandon the idea of a once-and-for-all critique, critique and reflection must then be described in terms of negotiation, and therefore an epistemology placed in a critical perspective becomes dialogical. It is illuminating to return to the situation that existed before Kant. As interpreted during the Enlightenment, as for instance by Bayle, a critique must be rooted in public debate and become a means for demonstrating misunderstanding and prejudice. Here we find the notion of critique imbedded in a social interpretation of negotiation. Kant claimed that critique itself has to be scrutinised, and that Reason has to address Reason in a pure self-critique. This turned critique into an inner-subjective enterprise rooted in the idea that the task of a philosophical epistemology is to find an unquestionable basis for knowledge. This separated critique from the public domain and eliminated any relationship between critique and negotiation. Accepting that critique cannot be a once-and-for-all task but must be an ongoing obligation, means that critique becomes dialogical, and this raises the original idea of the Enlightenment.

The possibility of knowledge conflict can be interpreted as a source of reflection, and knowledge conflict as an epistemic manifestation of critical situations which provide the ontological background for critique. Knowledge conflict provokes reconceptualisation. Therefore, a dialogical epistemology becomes a part of the epistemic framework for critical education. The step from a monological to a dialogical epistemology is a way of making impossible the notion of knowledge as authority and of education as 'delivery'.

5. KNOWING – AN EXPLOSIVE CONCEPT

With the assumption that knowing is an open concept having an affinity with promising, knowing became detached from any foundation, and critique became necessary whenever a knowledge claim was to be made; therefore, critique became an educational concept and the epistemic perspective became dialogical.[15] But as soon as negotiation and dialogue enter epistemology, knowing gets out of control. To think in terms of a dialogical epistemology means that knowing can no longer be thought of as an imprecise concept with an open texture. Even this would be to assume too much about the precision of the concept.

Because negotiation implies an interpersonal relationship, knowing becomes related not only to uncertainty and to prejudice but also to power.[16] We have touched upon the concept of power in different ways. From the thesis about linguistic relativism and the idea that mathematics can be seen as a language, we arrived at the notion of the formatting power of mathematics. But this notion can be developed further, because it does not make much sense to talk about the formatting power of mathematics in isolation. The notion of the power of mathematics makes sense only when understood as an interplay with other power structures of society, and this opens up further perspectives which transcend the present analysis. This means that if we really arrive at a concept of 'knowing' crucial to the analysis of critical mathematics education, then suddenly we find the analysis exploding in different directions. Not even a single and apparently specific notion such as 'knowing' can be understood unless we relate it to a variety of aspects which are not touched upon in my present analysis. We cannot round off the whole analysis of critical mathematics education unless we understand the concepts used in the analysis, and to understand a concept like 'knowing' presupposes analyses which are much more complex than the complete analysis itself. This is what I mean by an explosive concept.

A controlled concept can be fixed by a definition which contains it in a precise way. An open concept does not have fixed boundaries but its essential meaning can nonetheless be located somewhere in the landscape – it is possible to locate a hill, although it may be impossible to locate where it begins and where it ends. An explosive concept is different. When we make a detailed investigation of such a concept, shaking it and trying to open it, it may explode in different directions and carry us to places quite different from the original ground of our analysis.

I began this analysis by looking for an object of critique and by means of an analysis of the formatting power of mathematics, I concluded that because of mathematical modelling, mathematics has an important social influence; it follows that to understand this formatting power becomes an essential aspect of critical mathematics education. The object of critical activity was, therefore, located in the applications of mathematics. Further analyses brought me to see that this object of critique could be interpreted in a broader sense, and mathemacy was ascribed an importance similar to that of literacy, as a competence by means of which we become able to interpret and to understand features of our social reality. Mathemacy and social imagery became related through the notion of *Mündigkeit*. Then it was realised that we must also take into consideration the epistemic subject. Critique cannot be carried out as a forced activity. The learning subject must be the agent of critical activity, and the introduction of 'intentionality' gave a new perspective to the examples of the projects showing the importance of the students' understanding of the directions and the aims of the educational process. The students must become 'owners' of these aims in order to make the process a part of critical activity. It was also noted that this idea was incorporated in the examples which, therefore, could be re-examined with the critical subject in mind.

The complexity of the notion of critique could be further underlined by the following idea. It might well be impossible to make a distinction between a critical subject and a critical object, if an activity is to be critical at all.[17] Let me explain. Naturally I draw the distinction as an analytical line – and this had already been done several times during the previous analysis – but this distinction cannot have any ontological bearing in a situation which claims to be part of a critical education. This is because whenever a critique is addressed by a subject, this critique is also related to the pre-understanding of the subject and that pre-understanding always has to be included in a critique, both as part of the foundation of the critique and as part of the object of the critique. Therefore, the object of critique always has to include its subject, making the distinction partly a confusion.

This idea makes it possible to understand the nature of the Vico Paradox better: Human beings are unable to understand their own creations. I relate the Vico Paradox to the products, erected by means of our technological competence but which, nevertheless, we are unable to understand, and the functions of which we are unable to describe and to evaluate. Human creations have effects which cannot always be grasped by humanity. The Vico Paradox expresses the pessimism built into the philosophy of technology by identifying limits of knowing, but I still want to use the notion of 'reflective knowing' as referring to that part of our resources by means of which we try to struggle with the Vico Paradox. By means of reflection we try to grasp our situation as embodied in technology. Therefore, 'reflective knowing' reflects both a pessimism and an optimism: Reflections are necessary because of the frightening nature of technological development, and reflections are possible because we are not doomed to act blindly. This statement suggests the possibility of exercising power through technology and also the possibility of reacting to authority, meaning that empowerment also becomes a part of the conception of 'coming to know'. What we have created becomes the object of our critique as well as the basis for the critique, and the Vico Paradox is a reality.

This brings us to the *politics of knowing*. Knowledge is an expression of power, but not only that. Knowledge can also be a reaction to power. When knowing is combined with critique, knowing refers to crises which form the basis of critical activity. This places epistemology in a sociological and political context. To think in terms of the politics of knowing reveals that the educational perspective, which I have put on knowing, is far from being the only perspective which could be placed on it. My focus on critical education, and especially on critical mathematics education, provides only a partial and narrow look at a totality which the notion of 'politics of knowing' brings to light. I shall not say a single word about what that totality might encompass, except to say that Orwell, in *Nineteen Eighty-Four*, also addresses questions which I find crucial to a discussion of the politics of knowing.

NOTES

INTRODUCTION

[1] See Orwell (1987, pp. 312–313).
[2] See *Tractatus*: 3.323, 3.324 and 3.325.
[3] See *Tractatus*: 5.6.
[4] Normally, mathematics maintains a specific position in logical positivism. Mathematical statements are interpreted as tautologies and not excluded by references to the principle of verification.
[5] See Sapir (1925, p. 209).
[6] See Whorf (1956, p. 212).
[7] See Whorf (1956, p. 221).
[8] Normally, when I make a specific reference to primary school pupils I talk about 'children', secondary and upper secondary pupils I prefer to call 'students' and when speaking in general I also prefer to talk about 'students'. Sometimes, as I have already done in this note, I shall use 'pupil' instead of 'student'.
[9] Compare Glimm (ed.) (1991) with respect to this point.

CHAPTER 1

[1] 'Education after Auschwitz' ('Erziehung nach Auschwitz') is published in Adorno (1971). See also Paffrath (ed.) (1987).
[2] An excellent English introduction to Critical Theory is to be found in Held (1980). See also Benhabib (1986), Connerton (1980), Jay (1973) and Kellner (1989).
[3] See Habermas (1984, 1987).
[4] We could try to divide conflicts into different types, like: class exploitation, racism, sexism, elitism, exploitation of nature, struggles about resources, etc. Taking a different perspective we could enumerate personal crises, marital conflicts, family crises, problems at the workplace, etc. Such categorisations may be useful tools for drawing our attention in some direction, although it may cause an unjustified simplification. No categorisation can be firmly established but this is not essential to my argument here.
[5] See 'Proof of an External World' in Moore (1959).
[6] The term 'crisis' is thus interpreted in a broad sociological way, but this does not include every use of 'crisis'. I do not include, for instance, the interpretation of 'crisis' used when talking about the crisis in the foundation of mathematics at the beginning of this century.
[7] This formulation means a modification of the thesis of linguistic relativism mentioned in the Introduction. I do not think it possible to exclude 'crisis' from our linguistic perspective.
[8] Munir Fasheh has expressed the idea of conceptual blindness in the following way: "Hegemony is not only characterised by what it includes but also by what it excludes; by what it renders marginal, deems inferior and make invisible." (Fasheh, 1993).

⁹ This use of the term ideology is not the only one possible. Ideology could also mean a belief system of any kind. This interpretation has the immediate advantage that it does not refer to anything else but a belief system. The problem with the suggested, more restrictive use of the term is that it seems impossible to outline any general criteria for identifying ideological structures. If it is impossible to give any strict definition of a crisis, how could it be possible to maintain a distinction between a belief system which is able to hide a crisis and a belief system in general? The problem which becomes urgent is whether it is possible to find a distinction between 'ideology', 'belief system' and 'knowledge' in the same obvious way as the nature of the distinction between the conditions of the state of 'air', 'liquid' or 'solid'. Or will everything become mixed up?
¹⁰ This seems like a modification of the thesis of linguistic relativism. We have, however, to think of 'reality' as including ideological structures.
¹¹ The dissolution of the Soviet Union does not prove this point, because I do not conceive the Soviet Union as having anything to do with Marx's conception of a classless society.
¹² Habermas (1971) has outlined the knowledge-constitutive interest in emancipation as paralleling psychoanalytical therapy.
¹³ The notion of black dialectics has been drawn to my attention by Jens Højgaard Jensen.
¹⁴ In fact *The Principle of Hope* is the English title of a book by Ernst Block, in which he argues against the (short term) pessimism incorporated in traditional Marxism.
¹⁵ An overview of critical education is given in Hoffmann (1978). See also Anzinger and Rauch (eds.) (1972), Raith (ed.) (1973), and Tybl and Walter (eds.) (1973). For an English introduction see Young (1989).
¹⁶ The German word *Fachkritik* designates the activity of 'going behind' the curriculum and asks for the logical, sociological and political assumptions which constitute the subject matter as such.
¹⁷ See Negt (1964), Mollenhauer (1973), Lempert (1971) and Illeris (1974).
¹⁸ See for instance the discussion of the social reconstructionist movement, closely related to the work of John Dewey, in Giroux (1989).
¹⁹ See Freire (1972, 1974).
²⁰ See Wagenschein (1965, 1970).
²¹ Critical education was a main source of inspiration for educational development in Germany and Scandinavia during the 1970's. As an example of this, two universities were built in Denmark – Roskilde in 1972 and Aalborg in 1974 – which provide study programmes based on project work in all subjects (including mathematics). However, the broader influence of critical education has diluted the original educational enterprise. In many cases a pragmatic trend has blurred the original ideas of project organisation and problem orientation. What was once a radical renewal of education practice has, to a large extent, adjusted itself to fit conventional patterns in education and society.
²² The term *Lebenswelt* (lifeworld) is used by Habermas (1987) in his discussion on the extent of the expropriation of our daily life situation by the technological demands of systems and structures.
²³ How do I know this? How do I know that critical education has to take into consideration the critical background of schooling? My best answer would be: I do not see such statements as consequences of some other more general statements, they do not belong to some theory of critical education. I am not engaged in providing such a foundation. I try to make interpretations, and as I interpret 'critical education' my statements are supposed to make sense.
²⁴ Giroux (1989, p. 214).

[25] See Giroux (1989, p. 147).
[26] Giroux (1989, p. 148).
[27] Giroux (1989, p. 151).
[28] Giroux (1989, p. 152).
[29] Giroux (1989, p. 152).
[30] Another important modification is made by Giroux. To be literate does not mean to be free in all senses of the term. Literacy, as part of a critique of ideology, can create a degree of freedom to think and perceive independent of some ideological constraints (but definitely not free from all constraints), but freedom in this ideological sense is not freedom in a material sense. Freedom to think could still mean freedom to think whatever you want, inside a jail. As Giroux put it: Illiteracy does not explain the causes for mass unemployment, bureaucracy, or the growing racism in major countries like the U.S.A. Literacy does not automatically reveal, nor guarantee, social, political and economical freedom. However, literacy can mean a step towards empowerment. See Giroux (1989, p. 155).
[31] A similar notion, 'matheracy', has been developed by Ubiratan D'Ambrosio.
[32] Later we have to explore the role of mathematics in society to find a better foundation for the interpretation of mathemacy. See Chapter 3.

CHAPTER 2

[1] Dewey (1966, p. 87).
[2] In *Models of Democracy* David Held writes: "... nearly everyone today says they are democrats no matter whether their views are on the left, centre or right. Political regimes of all kinds in, for instance, Western Europe, the Eastern bloc and Latin America claim to be democracies. Yet, what each of these regimes says and does is radically different. Democracy seems to bestow an 'aura of legitimacy' on modern political life: rules, laws, policies and decisions appear justified and appropriate when they are 'democratic'." (Held, 1987, p. 1).
[3] A classical discussion in the analytical philosophy of 'democracy' is found in Benn and Peters (1959).
[4] Such an interpretation is for instance found in Chapter 11: 'Democracy and Education' in Peters (1966).
[5] See for instance Bowles (1983).
[6] Thucydides (1972, p. 145).
[7] Rousseau (1968, p. 110).
[8] Democracy need not be linked to the question of governing a country or community. In what follows I shall have a more general approach in mind, meaning that democracy will be seen to be a characteristic of governing some sort of organisation, be it a large or small community or a private or public organisation. However, I shall still refer to these as 'societies'.
[9] Rousseau (1968, p. 113).
[10] It is interesting to keep in mind the dangers mentioned by Rousseau. Does the existence of wealth corrupt both the rich and the poor? Do the citizens become slaves of others or slaves of opinion?
[11] A few conditions about the organisation of society are stressed by John Stuart Mill, but not to any great extent. Mill belongs to the liberal tradition. See Chapter 4: 'Under What Social Conditions Representative Government Is Inapplicable' in *Considerations on Representative Democracy* in Mill (1975).

¹² However, perhaps this exists only as a potential capacity because only a certain attitude will stress the importance of a democratic way of social control, and a democratic attitude is not of a common nature.

¹³ See Mill (1975, p. 214).

¹⁴ Mill discusses problems connected with the election of representatives, and he raises questions about the nature of democratic competence by asking: How could anybody represent others in a democracy? "If it is important that the electors should choose a representative more highly instructed than themselves, it is no less necessary that this wiser man should be responsible to them; in other words, they are the judges of the manner in which he fulfils his trust: and how are they to judge except by the standard of their own opinions? How are they even to select him in the first instance, but by the same standard?" (Mill, 1975, p. 326).

¹⁵ See Schumpeter (1987, p. 269).

¹⁶ Schumpeter (1987, p. 269).

¹⁷ See Schumpeter (1987, p. 273).

¹⁸ Bell (1980, p. 542).

¹⁹ Bell (1980, p. 543).

²⁰ Also this formulation is misleading, because the question does not simply concern the relationships between groups of people but concerns the structure of those relationships in the sense that an undemocratic trend can develop even if it is not actually supported by anybody.

²¹ It is naturally possible to describe a number of other problems of democracy in highly technological societies.

²² In *Considerations on Representative Democracy* Mill observes: "But any education which aims at making human beings other than machines, in the long run makes them claim to have control of their own actions." (Mill, 1975, p. 185).

²³ The expressions 'liberal education' might be rather misleading. In 'Liberal Education and the Nature of Knowledge' P. H. Hirst has made an analysis of 'liberal education', and here the term is primarily related to some analytically defined structures of knowledge, while *Allgemeinbildung* has a broader definition.

CHAPTER 3

¹ This phenomenon of formatting is referred to by Philip J. Davis and Reuben Hersh in *Descartes' Dream*. They make a distinction between mathematics used in descriptions, predictions and prescriptions. The prescriptive use refers to situations where the use of mathematics leads to some sort of human or technological action: "We are born into a world with so many instances of prescriptives mathematics in place that we are hardly aware of them, and, once they are pointed out, we can hardly imagine the world working without them. Our measurements of space and mass, our clocks and calendars, our plans for buildings and machines, our monetary system, are prescriptive mathematisations of great antiquity. To focus on more recent instances ... think of the income tax. This is an enormous mathematical structure superposed on an enormous pre-existing mathematical financial structure. ... In American society, there are plentiful examples of recent and recently reinstated prescriptive mathematisation: exam grades, IQs, life insurance, taking a number in a bake shop, lotteries, traffic lights ... telephone switching systems, credit cards, zip codes, proportional representation voting ... We have prescribed these systems, often for reasons known only to a few; they regulate and

alter our lives and characterise our civilisation. They create a description before the pattern itself exists." (Davis and Hersh, 1988, pp. 120–121).

Several other authors have discussed mathematics as part of a social context. Let me just make a reference to Bloor (1976), Keitel (1989), Keitel *et al.* (eds.) (1989), Otte (ed.) (1974), Restivo *et al.* (eds.) (1993), Tymoczko (ed.) (1986) and Wilder (1981).

[2] See Dessauer (1958, p. 184).
[3] See Stork (1977, p. 1).
[4] See, for instance, Gehlen (1980).
[5] We are witnessing a de-personalised and de-humanised conquering of our *Lebenswelt* (lifeworld). For a discussion of this phenomenon, see Habermas (1987).
[6] See Benjamin (1973, p. 248).
[7] A discussion of the different types of technology is to be found in Jensen and Skovsmose (1986).
[8] The Vico Paradox is mentioned in Jensen and Skovsmose (1986).
[9] Bell (1980, p. 500).
[10] These two concepts are used by Alfred Sohn-Rethel (1970) in an attempt to discuss the social origin of certain basic forms of thinking. I shall not try to expound upon his attempt to 'socialise' epistemology. I just find it useful to apply those concepts. The ideas elaborated in this section have been developed in cooperation with Christine Keitel and Ernst Kotzmann as part of the BACOMET-III project, see Keitel, Kotzmann and Skovsmose (1993). See also Keitel (1993).
[11] In Chapter 6 we shall return to a discussion of the National Product as a mathematical term.
[12] In this context I am not going to consider the origin of thinking abstraction, although this discussion is as important as the discussion of the way thinking abstractions become realised. This restriction in my analysis has only to do with my concern for the thesis about the formatting power of mathematics. The thesis which would focus the discussion on the origin of thinking abstractions has to do with the social roots of mathematics. A principal point in such an analysis would concern the way thinking abstractions may be distilled from realised abstractions, i.e. the way in which abstract thinking becomes an expression of some social conventions, or social stereotypes. That becomes an analysis of the worldly life of the souls in Popper's Third World, and also more in accordance with Sohn-Rehtel's project.
[13] See Curry (1951).
[14] Systems development makes up a part of the practice of computer science.
[15] Beck (1986) uses the term 'risk society'. That risks can be related to mathematical modelling is suggested by Boos-Bavnbek (1991). See also Boos-Bavnbek and Pate (1989a, 1989b).
[16] See Bourdieu (1991).
[17] Bourdieu (1991, p. 170).
[18] See Skovsmose (1988).
[19] Niss (1983, p. 248).
[20] See for instance Bishop (1988a), Bishop (ed.) (1988) and Høines and Mellin-Olsen (eds.) (1986).

CHAPTER 4

[1] This project has already been analysed in Skovsmose (1984).
[2] During the 1970's many introductions to project work have been published. One of

them is Berthelsen, Illeris and Poulsen (1977), which summarises some of the experiences gathered in Denmark at that time and which also became a main inspiration for further experimental work. It is however characteristic of the attitude towards mathematics education among progressive educators at that time that this broad introduction to project work only mentioned mathematics once (concerning a calculation having to do with the length of a pump rod).

[3] See for instance Damerow *et al.* (1974), Münzinger (ed.) (1977), Niss (1977), Riess (ed.) (1977), Volk (1975) and Volk (ed.) (1979); this later work contains a great number of references.

[4] I do not find the rationale of the work of IOWO, as described in "Five Years IOWO" for instance, closely related to the notion of critical mathematics education. Nevertheless, IOWO has been very important in redefining the whole concept of mathematics education and also in making possible an educational practice which expresses many features of a critical education. The same can be said about the work of Hans Freudenthal, who in a most powerful way, has expressed criticism of the structuralist movements of the 1960's and identified mathematics as a human activity, but has never shown any interest in Critical Theory itself.

[5] See Skovsmose (1980, 1981a, 1981b). In Skovsmose (1985) I give a summary of my previous work.

[6] Let me mention a few of my main inspirations from the 1970's. From Stieg Mellin-Olsen I have learnt that considerations about mathematics education must be related to a complexity of educational interpretation and that the educational discourse can be embedded in a broad sociological framework; see especially Mellin-Olsen (1977). Mogens Niss has always emphasised the social importance of mathematics, and from that assumption he has developed the discussion not only of the applications of mathematics but also of the possibilities for evaluating those applications. I have been much interested in the work of Dieter Volk who has integrated a critical interpretation of emancipation through mathematics education, and at the same time pointed to constructivism as having potential for understanding mathematics. Through discussions with Bent Christiansen, I have learnt the necessity for a careful criticism of educational theses as a precondition for developing those theses. From Tage Werner, I have come to understand the extension of the blind spots of any educational theory: whatever the way a theoretical systematisation is performed, essential aspects of an educational situation are ignored. This idea is one of the 'axioms' of this book. Naturally, I have learnt from many other mathematics educators, but I have also received inspiration for critical mathematics education from educators not concerned with mathematics.

[7] The richness of the second wave is documented by an annotated bibliography (Volmink *et al.*, 1994) which makes it obvious that it is impossible to do justice to the complexity of that wave. I must, however, mention Abraham and Bibby (1988), Borba (1990), D'Ambrosio (1980, 1981, 1985a, 1985b), Frankenstein (1987, 1989, 1990), Frankenstein and Powell (1989), Gerdes (1985, 1986, 1988), Hoffman and Powell (1989, 1990), Shan and Bailey (1991), Vithal (1992), and Volmink (1989, 1990). See also Noss *et al.* (eds.) (1990) and Julie *et al.* (eds.) (1993). The annotated bibliography also demonstrates that the second wave in critical mathematics education has been almost entirely unaware of the existence of the first wave.

An interesting position is taken up by the fashionable discussion of constructivism. As I see it, this discussion has served two main functions. On the one hand, it has placed the activity of the learner at the centre of an epistemic discussion which is essential to critical mathematics education; on the other hand, the discussion has a rather narrow focus on the development of mathematical competences and as a consequence, many aspects concerning the 'politics of knowing' have been neglected. Therefore, I see the orientation of the general discussion of constructivism as diverging from the main perspectives of critical mathematics education.

Many other writers have produced ideas relevant to the further development of critical mathematics education. Of special importance to me are Bishop (1990), Dowling (1991), Ernest (1991), Evans (1990), Lerman (1989) and Nickson (1992).

[8] In Chapter 11 the genetic epistemology of Piaget will be discussed further.

[9] In one sense it is not any more neutral *not* to touch upon sensitive questions. Not to challenge inequality is also a forceful political action.

[10] This phenomenon is discussed further in Chapter 10.

[11] Most interesting in this context is Wittenberg (1963).

[12] See Wagenschein (1965, p. 300). It may be tempting to compare this with G. W. Leibniz who maintains that the individual monad (the 'atom' in the metaphysics of Leibniz) contains the complexity of the world in itself.

[13] See Wagenschein (1965, p. 301).

[14] For the development of the ideas of the Pythagorean theorem as basis for geometry, see 'Das Exemplarische Lehren als fächerverbindendes Prinzip' in Wagenschein (1965).

[15] Thematic approaches and project work are also developed without any preconception of critical education.

CHAPTER 5

[1] The interviewer was Thue Ørberg.

[2] This paradox has been stated explicitly by Mogens Niss.

[3] The set of data was so complex and rich that all the concepts of statistics which are introduced during primary and secondary school in Denmark could come into use if the figures were going to be structured and investigated in detail

[4] With exceptions, as for instance in Workshop 5 having to do with measuring.

[5] Setting a scene has naturally to do with the thematic approach, as this idea is outlined in the introduction to Chapter 4.

[6] I do not maintain that the three ways are the only ones, just that we have exemplified three different types of scene-setting.

[7] In fact it is an open question as to what the meaning of '*the* meaning' could be in this context.

[8] A characteristic phenomenon is that students, when asked what they are doing in mathematics, just try to repeat what they have learned; they are not able to talk *about* the subject.

[9] This will be discussed further in Chapter 10.

[10] Here I shall not try to maintain a sharp distinction between semantics and pragmatics. Semantics has to do with the meaning of a term while pragmatics has to do with the situations in which to apply the term. This distinction may naturally be made rather fuzzy by the thesis that the meaning of a term is connected to the way it is used. Therefore, to condense, I shall talk only about multi-faceted semantics.

[11] Michel Foucault has used the expression 'the archaeology of knowledge' but I have not tried to related my work to his.

[12] Naturally it need not be the case that the children participating in "Constructions" would have been drawn into that work. For them the work finished as it did, also because of some pedagogical reasons. My suggestion is only to illustrate what could be the meaning of a mathematical archaeology.

CHAPTER 6

[1] This means that I do not find it possible to reduce 'reflection' and 'critical thinking' to 'logical awareness'. Attempts have been made to relate critical thinking in education to informal logic and to criticism 'inside the disciplines' (for a discussion, see McPeck (1990)). My approach, however, has a broader scope.

[2] We have described the relationship between reflective and technological knowledge from a semi-logical point of view, stressing the impossibility of decomposing reflective knowledge into smaller bits each belonging to technological knowledge. This irreducibility could also be discussed from a sociological point of view, showing how the institutions and research communities which uphold and produce technological knowledge are separated from the social institutions which could become the presenter of reflective knowledge. This institutional separation is reflected even in education.

This split reflects a duality between the sciences and the humanities. From an analytical and philosophical point of view, the duality has been neglected by logical positivism due to the nomination of one scientific method as paradigmatic and by the promotion of the thesis of the unity of sciences. The duality has been overlooked by the hermeneutic tradition by its 'forgetting' about the sciences in favour of the establishment of a respectable method of interpretation. Nevertheless, it becomes essential to overcome this duality, if the result of technological constructions are to become subjected to critical investigations.

[3] Hans Freudenthal's use of 'reflection' is different from mine. In *Revisiting Mathematics Education* he relates the concept to a 'level-raising' activity associated with mathematical knowing. Freudenthal's concept belongs to a theory of development of mathematical knowledge, while I intend the concept to be oriented outside mathematics.

[4] See Chapter 11 for a further discussion of 'knowledge'.

[5] Jensen (1980) has had an especially important role in the discussion of mathematical modelling in Denmark. For further discussion of mathematical modelling see Blum and Niss (1991), Blum *et al.* (eds.) (1989), Blum, Niss and Huntley (eds.) (1989), Niss, Blum and Huntley (eds.) (1991) and Lange *et al.* (eds.) (1993).

[6] For a more detailed discussion of this phenomenon, see Keitel, Kotzmann and Skovsmose (1993).

[7] To stipulate the existence of such a relationship creates another philosophical problem, similar to that which faces every theory of correspondence. What does it mean for conceptual entities to correspond to reality? The situation seems to be that we have begun our investigation of the effects of mathematical modelling by stumbling over a philosophical problem which is impossible to solve. Furthermore, my formulation can be misleading if linguistic relativism is assumed. This relativism makes it difficult to maintain a sharp distinction between language and reality.

Another question has to be mentioned. I have described a crisis as a feature of reality. Does this square with a position of linguistic relativism? I see the problem something like this: Crises are real in the sense that they cannot be interpreted so that they disappear, yet still reality can be structured by language.

[8] SMEC has been investigated by Kirsten Hermann and Mogens Niss (1982), with the intention of providing an example of real mathematical modelling accessible to upper secondary school. The basic idea has been that it is important for students to come to understand the nature of the applications of mathematics by studying a real example. This analysis is an exemplification of the introduction of real mathematical modelling of social significance into mathematics education. It is also an exemplification of mathematical archaeology.

[9] See Hanson (1958).

[10] My use of the term 'system development' is different from the use found in computer science; here 'systems development' refers to the total process of 'realising systems'.

[11] It is also possible to use the described activities of a modelling process to outline a route leading from real abstractions to thinking abstractions (when we let a mathematisation become the 'final' step of the process). This indicates that it is misleading to interpret the route from real abstractions to thinking abstractions as the inverse of the route from thinking abstractions to realised abstractions. The elements of a modelling process interact in a more chaotic way. Nevertheless, one important perspective of that chaos is to see how mathematics is used as an element in a process of design, and this means a focus on mathematical formatting.

[12] It is important to observe that different possibilities for power pressure can be exercised along the routes of the modelling process. Some basic features of social life are connected to the way in which society executes and controls the development of systems, and the way it lets some of the developed systems materialise as a form of social order.

[13] In this context Russell's Theory of Descriptions is important. It suggests how to make translations from natural language, building upon what Russell conceived of as an insufficient grammar, into a formal language which leaves no possibilities for ambiguities to occur.

[14] In fact Carnap (1937) added the idea that the language of science may be extensional, meaning that a language developed along the ideas of set-theory is adequate for science.

[15] In *Ludwig Wittgenstein: A Memoir*, Norman Malcolm quotes the following comment of Wittgenstein: "We have the idea of an ideal model or an ideal description of what one sees at any time. But no such ideal description exists. There are numerous sorts of things which we call 'descriptions' of what we see. They are all rough. And 'rough' does not mean 'approximation'. We have the mistaken idea that there is a certain exact description of what one sees at any given moment." (Malcolm, 1966, p. 50).

[16] In this case we can see that the systemic language has an origin in an extended mathematical modelling.

[17] It must be noted that even if it makes sense to consider mathematics as a language, it need not make sense to identify mathematics as a language.

[18] Some overlap exists. 'Problem identification' has been described as one of the activities of pointed modelling, but I choose to include it also as part of the context for the modelling process.

[19] This can be assumed when we have to do with pointed modelling, but if we have to do with extended modelling, a problem orientation cannot be assumed. In this case the mathematical language comes to express some fundamental abstractions built into a metaphysical system, and by doing so the abstractions become more remote and hidden but at the same time more embedded in the social situation. For instance, we all know about the money system, but nobody seems able to express the basic assumptions which underlay it.

[20] Lakatos (1976, p. 90).

[21] The described tasks of reflective knowing may serve as an illustration of the concept of reflecting, but let me just point out some shortcomings of what has been said and done so far. The outline has been developed with pointed modelling in mind, which means that we have not touched upon extended modelling. So, is the developed terminology applicable in general? I do not suppose it to be so, but some amplification may be helpful. An analysis of an extended application has to be developed as part of a broader cultural analysis. As an example of such an analysis, see Swetz (1987).

[22] See for instance Lave (1988).
[23] An important objection to the distinction between mathematical, technological and reflective knowing can be raised: This distinction presupposes too narrow a conception of mathematical knowing. This objection is quite right. Nevertheless, I find the distinction useful in pointing out the importance of integrating a mathematical competence into a richer context. I find that my too narrow conception of 'mathematical knowing' has a strength in describing what actually is developed by ordinary mathematical teaching. That descriptive 'strength' is naturally also a weakness of the education described. The broader competence I have been looking for, I call 'mathemacy'. However, I could also have designated this competence as 'mathematical knowing'. In that case the use of 'mathematical knowing' would have a normative force and a weak descriptive content.
[24] See Davis (1993).
[25] This means refusing any curriculum theory which focuses on well-established knowledge structures (as for instance that of P. H. Hirst). From such a perspective, it becomes legitimate to develop mathematical knowledge isolated from reflective knowing.
[26] This is not a problematic experience, even if it is sometimes conceived of as such, when mathematics is contextualised and it turns out that children can solve the problems without mathematics. Then the contextualising is not seen as useful, because it did not force the children to use mathematics. They do not realise that mathematics is an important tool. But this is only a consequence if contextualising is seen as a way of illustrating the necessity of mathematics. However, this has been the axiom behind most present attempts in contextualising mathematics.
[27] This need not have been possible in the actual context of "Economic Relationships in the World of a Child". The children's final letters indicate that the interest in the subject had come to an end.
[28] My formulations may indicate some relationships between a pointed application of mathematics and reflective knowing in an elementary mathematics education. But what is one to think about extended modelling? Perhaps this type of modelling is most relevant to the concept of critical citizenship, but do reflections on extended modelling make sense in elementary mathematical education?

CHAPTER 7

[1] Examples of the descriptions of families are found in the following section of this chapter. One possibility was to use the student's own families as a source of description, but that possibility could easily cause problems and uncertainty, so we preferred to let the Micro-Society be imaginary.
[2] It became obvious that the facility of the database influenced the student's way of handling the distribution. It became easier to think in terms of increments instead of proportionality.
[3] I shall return to this question in my comments in Chapter 9 to the project "Energy".
[4] For a further discussion of these coordinates of mathematical practice see Kitcher (1984).
[5] The discussion of the use of mathematics has especially been related to what we have called pointed modelling. Perhaps extended modelling, as exemplified by the mathematics used in setting up a semantical landscape for economical discussion, is very influential. Perhaps an essential aspect of mathematics is to provide analytical tools to be used in interpretations of certain phenomena, meaning that the importance of mathematics is found in its potential for creating syntactical and semantical structures for a

technological discourse. This means that a main social influence of mathematics is not to be found in the direct applications of mathematical models but in the linguistic dominance of the discourse of system development, i.e. of the systemic language. Therefore, examples of extended modelling may show an essential influence of mathematics on social life.

However, if this analytical use of mathematics is so decisive, what is the consequence of this from the perspective of education? If the analytical use of mathematics is essential and in fact exercises an important power by influencing and restructuring not only technological actions but also the basic language of technological discourse, what does this mean for attempts to establish critical mathematics education? If we were to try to illustrate this aspect of mathematics in education, in which direction do we then have to look? Furthermore, even if we come to see this aspect of mathematics as being essential (from the perspective of philosophy of science) it need not be essential as part of critical education. The analytical force of mathematics could be too 'intangible' to become a guideline for educational practice. In short, these considerations require a discussion related to a different educational project which sets out to illustrate the enchanting powers of the mathematical language.

[6] Iben Maj Christiansen from Aalborg University poses this question as the focus of her research. She investigates dialogue in the classroom from the perspective of critical mathematics education.

[7] From an educational point of view such a preoccupation may have an advantage, but it looks different when seen from the perspective of a technological action.

[8] The importance of challenging questions as part of a critical approach to mathematics education has been discussed by Morten Blomhøj.

[9] This idea shall be developed further in Chapter 11.

CHAPTER 8

[1] See Chapter 1, Section 5.

[2] Even if literacy and mathemacy provide a means for decoding social phenomena, they need not be the only competencies developed in school with such potential. And mathemacy could relate to other activities than mathematical modelling. For instance, mathemacy may have to do with an awareness more directly related to the educational situation in which the students are situated, indicated by questions like: Why do we have to learn things like this? Why are such exercises found in textbooks?

[3] Besides 'democratic competence', we may talk about a 'democratic attitude', meaning that it is not only of importance that the participants in a democracy uphold certain competencies, they must also maintain certain attitudes. "Our Community" illustrates that a conception of democracy can also be extended in the direction of 'attitude'.

[4] This aspect will be expanded in Chapter 10 which deals with intentionality.

CHAPTER 9

[1] Throughout this chapter I talk about 'use of energy' as usually spoken of in everyday language. Physics states that energy does not disappear but changes from one form to another. It is this phenomenon of changing which is referred to by the expression of 'use of energy'.

² The quality of food is not defined simply by means of kJ. Neither a large amount of kJ nor a small amount determines whether the breakfast is of good quality or not. Other factors like the fibre content have to be taken into account.

³ In fact it was not obvious to the students what 'area' meant in this case. In mathematics teaching, areas have to do with simple regular shapes, bounded by straight lines or regular curves.

⁴ The formulas are described in Bjerre (1985) as well as the possibility of estimating bike-resistance and the cyclist's use of energy. Clearly the formulas are simplifications, for instance the wind resistance also depends on the density of the air. The formulas comprise different types of resistance, the resistance from the wind and the resistance caused by the bike itself. The front area a has to be measured in m² and the velocity v in m/s. The formulas then gives the bike resistance r as measured in N (Newton).

⁵ The cycling power c is measured in kJ/h (which explains the factor 3.6). The cyclist's use of energy e is measured in kJ, and the time t in hours. The last formula shows that at rest we are also using energy which is estimated to be 400 kJ/h. The factor 4 indicates the efficiency in cycling; the cyclist is using energy in different ways (sweating, etc.), and only 25% of the used energy is transformed into 'pedal-energy'.

⁶ Much agricultural research has been directed to determine the figures necessary to calculate an 'energy account' of farming.

⁷ Naturally, the calculations would look different, if the more general use of petrol is also taken into consideration: driving to the market to buy seed, for instance. The figures have only to do with matters directly connected with the actual field work. Further, the 'energy account' looks different if we take into account the step from corn to the final result: a loaf of bread in the baker's shop. The energy used in farming makes up only about 20% of the total use of energy in bread production.

⁸ Also this calculation could be continued further, if the step from the bacon factory to the shops is also taken into account. Not to mention the energy used in cleaning the shops, keeping the freezer going, etc.

⁹ This point will be developed further in the next chapter concerning 'intentionality'.

¹⁰ I ignore the discussion of whether such a description in some cases can be exactly correct or if some sort of approximation is always necessary – this is not important now; I simply claim that in some cases mathematics has a strong descriptive power.

¹¹ This possibility makes qualitative methods necessary as autonomous scientific methods, and the same is the case for formal methods.

¹² Although the formulations so far have expressed a sort of symmetry between formal and non-formal descriptions this is not always the case. We have to address the following question: If we have to discuss the relationship between two different sorts of descriptions to find out which one is more reliable in a certain context – and let us say the descriptions are of a different nature, one using a formal and one a natural language – in what language is the comparison going to take place? In this situation natural language takes superiority, no formal methods exist by means of which we can argue that a certain description is to be preferred. The meta-language for the discussion and evaluation of descriptions is non-formal. However, the necessity of a non-formal meta-language does not imply that natural language descriptions must be preferred.

¹³ A further question, not illustrated by "Energy", is: What about 'classical' mathematical competence as being able to handle tricky mathematical problems? When we see "Energy" as an argument for interpreting the different competences defining mathemacy as being of importance, we have based the argument on the applications of mathematics in making the input–output figures, and this is a technological use of mathematics. Yet we could discuss mathematical 'ingenuity'. How do we interpret the preconditions of

such talents from the perspective of critical mathematical education? I have to admit that this also is one of the questions I shall not touch upon in this book.

[14] The sort of empowerment we have been talking about here is different from the sort which emerges as the self-confidence of a student who realises that he or she is able to do better of mathematics than the rest of the class. This sort of confidence is not what I have been concerned about, but sometimes this interpretation is mixed up with the idea that mathematics is empowering.

[15] Here a new analysis could begin. I find that a basic feature of a highly technological society is that new types of risk structure emerge. I find it useful to talk about a 'risk society'. A source of the risk structures I find in the transition between formalisation of language and the formalisation of routines. We base our technological management on investigations by means of formal structures. This means we perform an actual technological action by means of an investigation of a set of hypothetical situations. These situations are constructed by formal means, and often we have no opportunity to compare structures of the hypothetical situation with an actual situation. This means that acting involves taking a risk.

[16] If mathematics is empowering, we immediately meet the problem of elitism. Mathemacy was described as part of a democratic competence. (It has to be remembered that "Our Community" broadened our understanding of the position of mathemacy as part of a democratic competence.) If we admit that a particular talent for mathematics is essential, we easily confront elitism. It was emphasised by the early philosophers of democracy that a competence to judge what is done by the 'people in charge' does not presuppose a specific education. (Were that the case, democracy would seem to be substituted by 'expertocracy'.) However, if mathemacy plays a role in a critical citizenship, it becomes problematic whether that competence is described in ways which presuppose advanced teaching of mathematics, or presuppose specific talents in mathematics. In this analysis, I am not going to relate critical mathematics education to advanced mathematics teaching (although I conceive it as meaningful to talk about critical mathematics education also as part of a university study, but that has to be argued in a different way). I am considering elementary mathematics education, so the competence of mathematics being discussed is that which could be generally obtained. What we have to be aware of is the contradiction which might arise in developing mathematics education as part of a democratic force and at the same time, stressing the necessity to maintain an advanced mathematics.

[17] The possibility remains that the three mentioned alternatives are not exhaustive. For instance, it could be maintained that it is most essential that the students find the opportunity for action as an extension of the educational process. This means that what has to be developed in "Energy" is the last phase where the students tried to do something about the use of electricity at home and at school. This is the active part of the project, and this is what characterises the critical dimension.

CHAPTER 10

[1] If we were all just mechanical systems, fully predictable once all the parameters had been specified, would this imply that whatever we would normally call an action is in fact a predetermined behaviour? Would it imply that the idea of action is produced only by our imagination, perhaps due to the fact that we cannot discern the multiple causes which force us to do whatever we are doing? Is the consequence of this that the concept of 'responsibility' loses its meaning? Could Pinocchio, in the strings, be called

responsible? Further, if we deny the possibility of being complex clockworks, do we then have to accept that human beings share some 'spiritual' constituent, the activity of which is not explicable in mechanical terms? However, even though these questions are of philosophical significance, my discussion will not proceed in this direction. I maintain that it makes sense to talk about human actions, but also that in many cases we cannot talk about the behaviour of a person as being an action – for instance, when the person is doing something as a reflex or out of habit or because the person is forced to do something.

[2] Both those conditions need heavy modification. It need not be the case, strictly speaking, that the impossibility of choice makes actions impossible. Think of the figure from a novel of Jean Paul Sartre finally obtaining his (existential) freedom only when he ends up in jail. We could also have extended the analysis of choice further into the concept of power, arguing that the possibility of action presupposes the possession of power. By performing an action, power is exercised. The assumption of awareness also needs modification. It need not be the case that I am fully aware of my action, even if I still want to label it an action. Awareness can be analysed further and we could for instance use a distinction between consciousness, preconsciousness and unconsciousness.

[3] An important source for my reference to intention and action is Searle (1983).

[4] The interpretation of intentionality can be extended in different directions. I shall take the direction of philosophy, which means that 'mentally oriented' is used primarily here as a philosophical concept, not as a psychological one.

[5] A difficult question, which I am not going to consider in this context, is whether intentionality characterises a conscious being. This question is essential to the philosophy of mind, and an important step is to discuss whether intentionality can be described and reduced to other types of relationship.

[6] In many classical philosophical analyses of the concept of intentionality, a main idea has been that a person is aware of his or her intentions. This idea is not useful if the concept of 'intentionality' is to be used in a discussion of the concept of 'action'. Intention has to be assigned the possibility of being brought completely into view as well as of disappearing into the dark shadows of unconsciousness. Another question is whether it makes sense to talk about intentions which can never be expressed.

[7] An example of a dispositional concept is the 'fragility' of a pane of glass. That a pane is fragile does not mean that it will actually break. But if the pane is hit by a football it will easily break. A concept is dispositional if it describes an internal and potential tendency. For a discussion of dispositional concepts, see Ryle (1949).

[8] I have not used the concept of 'motivation'. One problem is that many analyses of motivation are based on different conceptual frameworks trying to reduce the term of intentionality into terms belonging, for instance, to biology.

[9] This fact could be interpreted from a purely logical point of view. If I am to be consistent, I am not free to maintain conflicting sorts of intentions. This limitation is similar to the demand of rationality in the theory of action, which tells us that a system of preferences has to fulfil some logical relationships, if a person's behaviour is to be called rational. I do not maintain that the structure of intentions has to make up any well-structured pattern; I only maintain that choosing some intentions makes it difficult or impossible to maintain certain other intentions.

[10] Christiansen and Walther (1986) explore activity theory as a basis for an interpretation of mathematics education, and it is emphasised that the aims of the students have an important role to play in the educational process. Mellin-Olsen (1987) emphasises 'activity' as a main educational concept: "Activity will refer to actions emerging from the individual's own motivation. Activity is related to the individual as a political

individual of society. This implies that the individual, as a member of society, is in a situation where he is permitted responsibility for his own life situation in particular and for society in general." (Mellin-Olsen, 1987, p. 30). And later: "Activities are thus about the decisions, projects and corresponding goals of the individual." (Mellin-Olsen, 1987, p. 36). Using activity theory Mellin-Olsen develops an educational position which exposes the political dimension of mathematics education. Glasersfeld (ed.) (1991) has also been of great importance to me.

A comment on my use of the term 'activity' is necessary. I do not use the word in the sense it is used in 'activity theory' which stresses that a goal is an important part of any activity. In activity theory the use of the concept 'activity' is similar to my use of 'action'. I use the word 'activity' as 'blind activity' with 'blind' emphasised. Using this terminology we may say that laissez-faire pedagogy leads to some sort of activity, but when the child's attention is drawn towards a specific subject in some accidental way, no action need be involved.

[11] Perhaps it is useful to distinguish between a learning activity and the accomplished learning seen as the result of the activity. It could be considered whether it is the learning activity or the accomplished learning which should be seen as the condition for the fulfilment of the intentions of the learning. In my interpretation it is the learning activity, although serious objections could be raised about this interpretation.

[12] Some of these matters have been raised by Nickson (1993). The fact that the dispositions of the learner influence the teaching–learning situation and also the specific communication between the teacher and the students, has been discussed in more detail in relation to a project directed by Ib Trankjær at Nyvangskolen in Randers. The theme of this project is the social features of the mathematical classroom, and Bjarne Würtz Andersen and Ane Marie Krogshede Nielsen, he has produced a rich variety of video material of classroom situations, which can be used as basis for further studies.

Together with Helle Alrø I have tried to interpret a few of the videotaped situations, and in some cases the communication between teacher and students seems best interpreted in a framework of learning as action (see Alrø and Skovsmose (1993)). The students try to find out what the meaning of a specific teaching–learning situation could be: – What are we aiming at? But because the students' questions have been well disguised, and sometimes even formulated as answers to the teacher's questions, they also become misinterpreted, and taken by the teacher to be 'normal' answers to 'normal' questions. In this way the material produced at Nyvangskolen has been a most useful empirical support for developing the theoretical framework described in this chapter. Finally, I must emphasise that the video material can be used in many other ways in order to facilitate discussion of mathematics education.

[13] See the works of Lena Lindenskov.

[14] The results found by Kirsten Grønbæk Hansen point in the same direction.

[15] It might be interesting to take a look back at the first section of Chapter 5, telling about students' opinions about mathematics.

[16] I do not use 'assimilation' with any reference to Piaget's discussion of assimilation and accommodation.

[17] When criticising genetic epistemology I do not include its interpretation of the early intellectual development of children; in this case, it may make more sense. What I have in mind is the perspective placed on the teaching–learning situation at school level. In Chapter 11 additional aspects of the genetic epistemology will be discussed.

[18] Wagenschein has anticipated some of those ideas about the involvement of the student, see Chapter 4, Section 5.

[19] See Skovsmose (1981b).

[20] Initially, I maintained that a condition for action is a degree of indeterminism, but later I asked whether it could be possible to act in a situation without possibility of any choice. Let the emergence of underground intentions be my ambiguous answer to this question.

[21] See Mellin-Olsen (1981).

[22] An astonishing example of (a sort of) instrumentalism is found in *A Mathematician's Apology* where G. H. Hardy writes: "I do not remember having felt, as a boy, any passion for mathematics, and such notions as I may have had of the career of a mathematician were far from noble. I thought of mathematics in terms of examinations and scholarships: I wanted to beat the other boys, and this seemed to be the way in which I could do so most decisively." (Hardy, 1967, p. 144). And later: "I had of course found at school, as every future mathematician does, that I could often do things much better than my teachers; and even at Cambridge I found, though naturally much less frequently, that I could sometimes do things better than the college lecturers. But I was really quite ignorant ... of the subjects on which I have spent the rest of my life; and I still thought of mathematics as essentially a competitive subject." (Hardy, 1967, pp. 146–147). These quotations only serve as an illustration, because I am unhappy about making far-reaching conclusions from 'extreme' examples. However, the example of Hardy indicates that some students may develop 'some sort of knowledge' from an instrumental attitude.

[23] See D'Ambrosio (1985b, p. 472).

[24] This links up with the observations made in relation to the project "Economical Relationships in the World of a Child", see Section 3 in Chapter 4.

CHAPTER 11

[1] This last assumption is, for instance, formulated by Wittgenstein in the *Tractatus*, where it is maintained that an arbitrary proposition is a truth-function of elementary propositions, not able to constitute contradictions. Every elementary proposition states a simple fact, and the facts make up the world.

[2] See the dialogue *Theaetetus*.

[3] An important example of an interpretation of education as being associated to a certain body of knowledge is found in the analysis of P. H. Hirst. Hirst, however, does not assume the existence of a single integrated body of knowledge but specifies different knowledge areas, each of which must have a position in a curriculum. Nevertheless, I see Hirst's position as being in line with the claim of the existence of an authorised body of knowledge, because the different areas of knowledge, as set up by Hirst, do not contradict each other. They have divided the land of knowledge between them and seem to be living peacefully side by side. And united, they make up an authorised body of knowledge.

[4] However, not to combine knowledge with truth immediately leads to difficulties. If knowledge is not connected to truth, we have introduced relativism. And how can we stop this? If every statement can become part of some person's knowledge, if he or she believes a statement and has sufficient reason to maintain it, then knowledge easily becomes 'relativised' to such a degree that we are unable to separate 'belief' and 'knowledge'. The problem is how to step out of the classical definition of knowledge without falling involuntarily into an empty definition and wiping out every difference between 'knowledge' and 'belief'.

[5] It must be noted that 'the truth of p' need not be described by a corresponding theory

of truth, as done above. Other conceptions of truth could be assumed, but it will not make any difference in this context.

[6] In the paper 'What Is a Speech Act', Searle has made an analysis of 'promise', which has been most helpful to me. The possibility of a non-descriptive analysis of 'knowledge' was introduced by Austin in the article 'Other Minds'. Here he relates 'knowledge' and 'promise'.

[7] I find Wittgenstein's discussion of the private language problem important in this context. Private language is impossible, and so is private knowledge, and for similar reasons. (So what about Robinson Crusoe? Was it really impossible for him to know anything?)

[8] The notion that knowledge comprises a non-personal homogeneous body of truths has been attacked from a relativistic position which maintains that the truth of a proposition always depends on the context in which it is uttered, and, further, that it is impossible to decide which context is right for determining the truth value of the proposition. According to relativism, it does not make sense to state 'p is true'; we have to state 'p is true in the context c'. Next, we have to face the possibility that the two propositions 'p is true in context c' and p is false in context d' are both true. I agree with this conclusion. However, I do not agree when we come to the basic axiom in the relativistic position: All contexts are equal. Nor do I like the modified relativism which says that a proposition is true, if it is held true by the majority. To me the problem is that relativism eliminates the difference between 'knowledge' and 'belief'. I find that two different promises are offered by the declarations: 'I know p' and 'I believe p'. I hope that the given interpretation of knowledge does not collapse into relativism, but at least it does not repeat the absolutism built into the classical definition.

[9] In the hermeneutic tradition of Hans-Georg Gadamer we find a positive interpretation of terms like prejudgement, tradition and authority, see Gadamer (1989, pp. 277ff.). However, I find that the concept of critique becomes a necessity if we are not to let hermeneutics degenerate into conservatism.

[10] In Piaget's formulation: "The essential fact is that ... logico-mathematical experiences have to do with the actions which the subject carries out on the objects." (Beth and Piaget, 1966, p. 232).

[11] See Beth and Piaget (1966, p. 308).

[12] This line of thought has been elaborated in activity theory, but without accepting the Piagetian epistemology in general. See for instance Christiansen and Walther (1986).

[13] My use of 'reflective knowing' is quite different from 'reflective abstraction'. Reflective knowing has the use of mathematical competence as its object. It is a way of stepping outside the cathedral of formal knowledge to take a more general view of it. With reflective abstractions we continue to be placed inside the same body of knowledge. The whole interest of Piaget's epistemology is the development of mathematical knowledge itself. It is not part of Piaget's epistemology that mathematical knowledge or its applications have to be objects for reflection. But the importance of reflective knowing is precisely that the applications of mathematical knowledge are critical, and this is not conceptualised within Piaget's epistemology.

[14] A and B need not literally to be interpreted as persons. They can be seen as traditions (in a hermeneutic interpretation), or as paradigms (in a Kuhnean interpretation), or as texts, in addition to being interpreted as persons.

[15] See also Brousseau and Otte (1991).

[16] As mentioned the 'agents' of the interpersonal relationship might be 'traditions' or paradigms as well as persons.

[17] This important idea was suggested to me by Iben Maj Christiansen.

BIBLIOGRAPHY

This bibliography is selective in the following way: I have included the works in English which have been of relevance to my present study; with respect to the work in German I have been much more selective; of Scandinavian literature I have included only a few titles, although a major basis for my work has to be found here. Some of the German literature I have studied in Danish translation but in this bibliography I refer to the original editions.

Abraham, J. and Bibby, N. (1988): 'Mathematics and Society: Ethnomathematics and the Public Educator Curriculum', *For the Learning of Mathematics* 8(2), 2–11.
Adorno, T. W. (1971): *Erziehung zur Mündigkeit*, Suhrkamp, Frankfurt am Main.
Alrø, H. and Skovsmose, O. (1993): 'Det var ikke meningen! Om kommunikation i matematikundervisningen', *Nordic Studies in Mathematics Education* 1(2), 6–29.
Anzinger, W. and Rauch, E. (eds.) (1972): *Wörterbuch Kritische Erziehung*, Raith Verlag, Starnberg.
Apple, M. W. (1982): *Education and Power*, Routledge and Kegan Paul, London.
Austin, J. L. (1946): 'Other Minds', *Aristotelian Society* 20, 148–187. (Reprinted in Austin (1970), 76–116.)
Austin, J. L. (1962): *How to Do Things with Words*, Oxford University Press, Oxford.
Austin, J. L. (1970): *Philosophical Papers*, 2nd edition, Oxford University Press, Oxford.
Ayer, A. J. (ed.) (1959): *Logical Positivism*, The Free Press, New York.
Beck, U. (1986): *Risikogesellschaft: Auf dem Weg in eine andere Moderne*, Suhrkamp, Frankfurt am Main.
Bell, D. (1980): 'The Social Framework of the Information Society', in Forrester, T. (ed.): *The Microelectronics Revolution*, Blackwell, Oxford, 500–549.
Benhabib, S. (1986): *Critique, Norm, and Utopia: A study of the Foundations of Critical Theory*, Columbia University Press, New York.
Benjamin, W. (1973): *Illuminations*, Edited and with an Introduction by Hanna Arenth, Fontana Press, London.
Benn, S. I. and Peters, P. S. (1959): *Social Principles and the Democratic State*, George Allen and Unwin, London.
Berthelsen, J., Illeris, K. and Poulsen, S. C. (1977): *Projektarbejde*, Borgen, Copenhagen.
Beth, E. W. and Piaget, J. (1966): *Mathematics, Epistemology and Psychology*, Reidel Publishing Company, Dordrecht.
Bishop, A. J. (1988a): *Mathematical Enculturation: A Cultural Perspective on Mathematics Education*, Kluwer Academic Publishers, Dordrecht.

Bishop, A. J. (1988b): 'Mathematics Education in its Cultural Context', *Educational Studies in Mathematics* 19, 179–191.
Bishop, A. J. (ed.) (1988): *Mathematics Education and Culture*, Kluwer Academic Publishers, Dordrecht. (Reprinted from *Educational Studies in Mathematics* 19(2), 1988.)
Bishop, A. J. (1990): 'Western Mathematics: The Secret Weapon of Cultural Imperialism', *Race and Class* 32(2), 51–65.
Bishop, A. J. (1991): 'Mathematical Values in the Teaching Process', in Bishop, Mellin-Olsen and Dormolen (eds.), 195–214.
Bishop, A. J. (1992): 'International Perspectives on Research in Mathematics Education', in Grouws (ed.), 710–723.
Bishop, A. J., Mellin-Olsen, S. and Dormolen, J. (eds.) (1991): *Mathematical Knowledge: Its Growth Through Teaching*, Kluwer Academic Publishers, Dordrecht.
Bjerre, O. L. (1985): *Cykelmatematik*, Danmarks Radio, Copenhagen.
Blomhøj, M. (1992a): 'Modellering i den elementære matematikundervisning – et didaktisk problemfelt', Tekst MI 58, Department of Mathematics, Royal Danish School of Educational Studies, Copenhagen.
Blomhøj, M. (1992b): 'Samspil mellem teori og praksis – en forskningspraksis i matematikkens didaktik', Tekst MI 59, Department of Mathematics, Royal Danish School of Educational Studies, Copenhagen.
Blomhøj, M. (1993): 'Modellerings betydning for tilegnelsen af matematiske begreber', *Nordic Studies in Mathematics Education* 1(1), 16–37.
Bloor, D. (1976): *Knowledge and Social Imagery*, Routledge and Kegan Paul, London.
Blum, W. et al. (eds.) (1989): *Applications and Modelling in Learning and Teaching Mathematics*, Ellis Horwood, Chichester.
Blum, W., Niss, M. and Huntley, I. (eds.) (1989): *Modelling, Applications and Applied Problem Solving*, Ellis Horwood, Chichester.
Blum, W. and Niss, M. (1991): 'Applied Mathematical Problem Solving, Modelling, Applications, and Links to Other Subjects – State, Trends and Issues in Mathematics Instruction', *Educational Studies in Mathematics* 22, 37–68.
Boos-Bavnbek, B. (1991): 'Against Ill-founded, Irresponsible Modelling', in Niss, Blum and Huntley (eds.), 70–82.
Boos-Bavnbek, B. and Pate, G. (1989a): 'Information Technology and Mathematical Modelling, the Software Crisis and Educational Consequences', *Zentralblatt für Didaktik der Mathematik* 89(5), 167–175.
Boos-Bavnbek, B. and Pate, G. (1989b): 'Expanding Risk in Technological Society Through Progress in Mathematical Modelling', in Keitel et al. (eds.), 75–78.
Borba M. C. (1990): 'Ethnomathematics in Education', *For the Learning of Mathematics* 10(1), 39–43.
Bourdieu, P. (1991): *Language and Symbolic Power*, Edited and Introduced by John B. Thompson, Polity Press, Cambridge.
Bowles, S. (1983): 'Unequal Education and the Reproduction of the Social Division of Labour', in Cosin, B. and Hales, M. (eds.): *Education, Policy and Society*, Routledge and Kegan Paul, London, 27–50.
Brousseau, G. and Otte, M. (1991): 'The Fragility of Knowledge', in Bishop, Mellin-Olsen and Dormolen (eds.), 13–36.
Bruner, J. S. (1960): *The Process of Education*, Harvard University Press, Cambridge, Mass.
Carnap, R. (1937): *The Logical Syntax of Language*, Routledge and Kegan Paul, London. (Original German edition from 1934.)

Carnap, R. (1959): 'The Old and the New Logic', in Ayer (ed.), 133–146. (Original German version 1930–1931.)
Carnap, R. (1959): 'The Elimination of Metaphysics Through Logical Analysis of Language', in Ayer (ed.), 60–81. (Original German version 1932.)
Christiansen, B. and Walther, G. (1986): 'Task and Activity', in Christiansen, Howson and Otte, 243–307.
Christiansen, B., Howson, A. G. and Otte, M. (eds.) (1986): *Perspectives on Mathematics Education*, Reidel Publishing Company, Dordrecht.
Connerton, P. (1980): *The Tragedy of Enlightenment*, Cambridge University Press, Cambridge.
Curry, H. B. (1951): *Outlines of a Formalist Philosophy of Mathematics*, North-Holland Publishing Company, Amsterdam.
D'Ambrosio, U. (1980): 'Mathematics and Society: Some Historical Considerations Pedagogical Implications', *International Journal of Mathematical Education in Science and Technology* 11(4), 479–488.
D'Ambrosio, U. (1981): 'Uniting Reality and Action: A Holistic Approach to Mathematics Education', in Steen, L. A. and Albers, D. J. (eds.): *Teaching Teachers, Teaching Students*, Birkhäuser, Boston, 33–42.
D'Ambrosio, U. (1985a): 'Ethnomathematics and Its Place in the History and Pedagogy of Mathematics', *For the Learning of Mathematics* 5(1), 44–48.
D'Ambrosio, U. (1985b): 'Mathematics Education in a Cultural Setting', *International Journal of Mathematical Education in Science and Technology* 16, 469–477.
Damerow, P. et al. (1974): *Elementarmathematik: Lernen für die Praxis*, Ernst Klett Verlag, Stuttgart.
Davis, C. (1989): 'A Hippocratic Oath of Mathematicians', in Keitel *et al.* (eds.), 44–47.
Davis, P. J. (1993): 'Applied Mathematics as Social Contract', in Restivo *et al.* (eds.), 182–194.
Davis, P. J. and Hersh, R. (1981): *The Mathematical Experience*, Birkhäuser, Boston.
Davis, P. J. and Hersh, R. (1988): *Descartes' Dream: The World According to Mathematics*, Penguin Books, London.
Dessauer, F. (1958): *Streit um die Technik*, Verlag Josef Knecht, Frankfurt am Main.
Dewey, J. (1966): *Democracy and Education*, The Free Press, New York. (First edition 1916.)
Dowling, P. (1991): 'The Contextualizing of Mathematics: Towards a Theoretical Map', in Harris, M. (ed.), 93–120.
Ellul, J. (1964): *The Technological Society*, Random House, New York. (First French edition 1954.)
Ernest, P. (ed.) (1989): *Mathematics Teaching: The State of the Art*, The Falmer Press, London.
Ernest, P. (1991): *The Philosophy of Mathematics Education*, The Falmer Press, London.
Evans, J. (1990): 'Mathematical Learning and the Discourse of Critical Citizenship', in Noss *et al.* (eds.), 93–95.
Fasheh, M. (1993): 'From a Dogmatic Ready-Answer Approach to Teaching Math Towards a Community-Building, Process-Oriented Approach', in *PDME 2: Political Dimensions of Mathematics Education*, Pre-Conference Papers, Johannesburg.
Fischer, R. and Malle, G. (1985): *Mensch und Mathematik*, Bibliographisches Institut, Mannheim.
'Five Years IOWO' (1976): *Educational Studies in Mathematics* 7(3).
Frankenstein, M. (1983): 'Critical Mathematics Education: An Application of Paulo Freire's Epistemology', *Journal of Education* 165(4), 315–339. (Reprinted in Shor,

I. (ed.): *Freire for the Classroom*, Boyton and Cook Publishers, Porthmouth, New Hampshire, 1987, 180–210.)
Frankenstein, M. (1989): *Relearning Mathematics: A Different Third R – Radical Maths*, Free Association Books, London.
Frankenstein, M. (1990): 'Critical Mathematical Literacy', in Noss *et al.* (eds.), 106–113.
Frankenstein, M. and Powell, A. B. (1989): 'Mathematics Education and Society: Empowering Non-Traditional Students', in Keitel *et al.* (eds.), 157–160.
Freire, P. (1972): *Pedagogy of the Oppressed*, Herder and Herder, New York.
Freire, P. (1974): *Cultural Action for Freedom*, Penguin Books, London.
Freudenthal, H. (1973): *Mathematics as an Educational Task*, Reidel Publishing Company, Dordrecht.
Freudenthal, H. (1978): *Weeding and Sowing*, Reidel Publishing Company, Dordrecht.
Freudenthal, H. (1991): *Revisiting Mathematics Education: China Lectures*, Kluwer Academic Publishers, Dordrecht.
Gadamer H.-G. (1989): *Truth and Method*, 2nd revised edition, Sheed and Ward, London. (First German edition 1960.)
Gehlen, A. (1980): *Man in the Age of Technology*, Columbia University Press, New York. (First German edition 1957.)
Gerdes, P. (1985): 'Conditions and Strategies for Emancipatory Mathematics Education in Undeveloped Countries', *For the Learning of Mathematics* 5(1), 15–20.
Gerdes, P. (1986): 'On Culture: Mathematics and Curriculum Development in Mozambique', in Mellin-Olsen and Høines (1986), 15–41.
Gerdes, P. (1988): 'On Culture, Geometrical Thinking and Mathematics Education', *Educational Studies in Mathematics* 19, 137–162.
Giroux, H. A. (1989): *Schooling for Democracy: Critical Pedagogy in the Modern Age*, Routledge, London.
Glasersfeld. E. v. (ed.) (1991): *Radical Constructivism in Mathematics Education*, Kluwer Academic Publishers, Dordrecht.
Glimm, J. G. (ed.) (1991): *Mathematical Science, Technology, and Economic Competitiveness*, National Academy Press, Washington DC.
Grouws, D. A. (ed.) (1992): *Handbook of Research on Mathematics Teaching and Learning*, MacMillan Publishing Company, New York.
Grønbæk Hansen, K. (1991): 'Matematikken i erhvervsuddannnelserne', *Psyke og Logos* 12(2), 409–421.
Habermas, J. (1968): *Technik und Wissenschaft als 'Ideologie'*, Suhrkamp, Frankfurt am Main.
Habermas, J. (1971): *Knowledge and Human Interests*, Beacon Press, Boston. (First German edition 1968.)
Habermas, J. (1976): *Legitimation Crisis*, Heinemann, London. (First German edition 1973.)
Habermas, J. (1984, 1987): *The Theory of Communicative Action I–II*, Heinemann, London and Polity Press, Cambridge. (First German edition 1981.)
Hanson, N. R. (1958): *Patterns of Discovery*, Cambridge University Press, Cambridge.
Hardy, G. H. (1967): *A Mathematician's Apology*, With a foreword by C. P. Snow, Cambridge University Press, Cambridge. (First edition 1940.)
Harris, M. (ed.) (1991): *Schools, Mathematics and Work*, The Falmer Press, London.
Held, D. (1980): *Introduction to Critical Theory*, Hutchinson, London.
Held, D. (1987): *Models of Democracy*, Polity Press, Cambridge.
Hermann, K. and Niss, M. (1982): *Beskæftigelsesmodellen i SMEC III*, Nyt Nordisk

Forlag Arnold Busck, Copenhagen.
Hilbert, D. (1968): *Grundlagen der Geometrie*, Teubner, Stuttgart. (First edition 1899.)
Hilbert, D. (1964): *Hilbertiana: Fünf Aufsätze von David Hilbert*, Wissenschaftliche Buchgesellschaft, Darmstadt.
Hirst, P. H. (1969): 'The Logic of the Curriculum', *Journal of Curriculum Studies* 1, 142–158. (Reprinted in Hooper, R. (ed.) (1971): *The Curriculum, Context, Design and Development*, Oliver and Boyd, Edinburgh, 232–250.)
Hirst, P. H. (1973): 'Liberal Education and the Nature of Knowledge', in Peters, R. S. (ed.): *Philosophy of Education*, Oxford University Press, Oxford, 87–111. (First published 1965.)
Hoffman, M. R. and Powell, A. B. (1989): 'Mathematics and Commentary Writing: Vehicles for Student Reflection and Empowerment', in Keitel *et al.* (eds.), 131–133.
Hoffman, M. R. and Powell A. B. (1990): 'Gattegno and Freire: A Model for Teaching Mathematically Unprepared, Working-Class Students', in Noss *et al.* (eds.), 205–215.
Hoffmann, D. (1978): *Kritische Erziehungswissenschaft*, Kohlhammer, Stuttgart.
Horkheimer, M. (1970): *Traditionelle und kritische Theorie*, Fischer Taschenbuch Verlag, Frankfurt am Main.
Høines, M. J. and Mellin-Olsen, S. (eds.) (1986): *Mathematics and Culture*, Casper Publ., Rådal, Norway.
Illeris, K. (1974): *Problemorientering og deltagerstyring*, Munksgaard, Copenhagen.
Jay, M. (1973): *The Dialectical Imagination: A History of the Frankfurt School and the Institute of Social research 1923–50*, Heinemann, London.
Jensen, J. H. (1980): 'Matematiske modeller – vejledning eller vildledning?', *Naturkampen* (18), 14–22.
Jensen, H. S. and Skovsmose, O. (1986): *Teknologikritik*, Systime, Herning.
Julie, C., Angelis, D. and Davis, Z. (eds.) (1993): *Political Dimensions of Mathematics Education: Curriculum Reconstruction for Society in Transition*, Maskew Miller Longman, Cape Town.
Kant, I. (1933): *Critique of Pure Reason*, Translated by Norman Kemp Smith, MacMillan, London. (First German edition 1781.)
Kapp, E. (1978): *Grundlinien einer Philosophie der Technik*, Stern-Verlag Janssen and Co., Düsseldorf. (First German edition 1877.)
Keitel, C. (1989): 'Mathematics and Technology', *For the Learning of Mathematics* 9(1), 7–13.
Keitel, C. (1993): 'Implicit Mathematical Models in Social Practice and Explicit Mathematics Teaching by Applications', in Lange *et al.* (eds) (1993), 19–30.
Keitel, C. *et al.* (eds.) (1989): *Mathematics, Education and Society*, UNESCO, Division of Science, Technical and Environmental Education, Paris.
Keitel, C., Kotzmann, E. and Skovsmose, O. (1993): 'Beyond the Tunnel-Vision', in Keitel, C. and Ruthven, K. (eds.): *Learning from Computers: Mathematics Education and Technology*, Springer-Verlag, Berlin, 243–279.
Kellner, D. (1989): *Critical Theory, Marxism and Modernity*, Polity Press, Cambridge.
Kitcher, P. (1984): *The Nature of Mathematical Knowledge*, Oxford University Press, Oxford.
Kuhn, T. S. (1970): *The Structure of Scientific Revolutions*, University of Chicago Press, Chicago (First edition 1962.)
Lakatos, I. (1976): *Proofs and Refutations*, Cambridge University Press, Cambridge.
Lakatos, I. (1978): *Mathematics, Science and Epistemology: Philosophical Papers 2*, Edited by John Worrall and Gregory Currie, Cambridge University Press, Cam-

bridge.
Lange, J. de et al. (eds.) (1993): *Innovation in Maths Education by Modelling and Applications*, Ellis Horwood, Chichester.
Lave, J. (1988): *Cognition in Practice*, Cambridge University Press, Cambridge.
Lempert, W. (1971): *Leistungsprincip und Emancipation*, Suhrkamp Verlag, Frankfurt am Main.
Lerman, S. (1989): 'Investigations: Where to Now', in Ernest (ed.), 73–82.
Lindenskov, L. (1990a): 'The Pupil's Curriculum: Rationales of Learning and Learning-Stories', in Noss et al. (eds.), 165–172.
Lindenskov, L. (1990b): 'Mathematical Concept Formation in the Individual', in Booker, G. et al. (eds.): *Proceedings of the Fourteenth Psychology of Mathematics Education Conference I*, Mexico, 61–68.
Lindenskov, L. (1991): 'Everyday Knowledge in Studies of Teaching and Learning Mathematics in School', in Furinghetti, F. (ed.): *Proceedings of the Fifthteenth Psychology of Mathematics Education Conference II*, Assissi, 325–333.
Lindenskov, L. (1993a): 'Exploring the Student's Own Mathematics Curriculum', in Malone, J. A. and Taylor, P. C. S. (eds.): *Conctructivist Interpretations of Teaching and Learning Mathematics*, CURTIN, University of Technology, Perth, 149–156.
Lindenskov, L. (1993b): 'Hverdagsviden og Matematik: Læreprocesser i skolen', IMFUFA, Roskilde University, Roskilde.
Malcolm, N. (1966): *Ludwig Wittgenstein: A Memoir*, Oxford University Press, Oxford.
Marcuse, H. (1964): *One-Dimensional Man*, Routledge and Kegan Paul, London.
McPeck, J. E. (1990): *Teaching Critical Thinking*, Routledge, London.
Mellin-Olsen, S. (1977): *Indlæring som social process*, Rhodos, Copenhagen.
Mellin-Olsen, S. (1981): 'Instrumentalism as an Educational Concept', *Educational Studies in Mathematics* 12, 351–367.
Mellin-Olsen, S. (1987): *The Politics of Mathematics Education*, Reidel Publishing Company, Dordrecht.
Mill, J. S. (1975): *Three Essays on Liberty, Representative Government and the Subjection of Women*, With an introduction by Richard Wollheim, Oxford University Press, Oxford. (The three essays were first published in 1861, 1861 and 1869.)
Mills, C. W. (1959): *The Sociological Imagination*, Oxford University Press, New York.
Mollenhauer, K. (1973): *Erziehung und Emanzipation*, Juventa Verlag, München.
Moore, G. (1959): *Philosophical Papers*, Unwin, London.
Münzinger, W. (ed.) (1977): *Projektorientierter Mathematikunterricht*, Urban und Schwarzenberg, München.
Negt, O. (1964): *Soziologische Phantasie und exemplarisches Lernen*, Europäische Verlagsanstalt, Frankfurt am Main.
Nickson, M. (1992): 'The Culture of the Mathematics Classroom: An Unknown Quantity?', in Grouws (ed.), 101–114.
Niss, M. (1977): 'The "Crises" in Mathematics Instruction and a new Teacher Education at Grammar School Level', *International Journal of Mathematical Education in Science and Technology* 8, 303–321.
Niss, M. (1983): 'Considerations and Experiences Concerning Integrated Courses in Mathematics and Other Subjects', in M. Zweng et al. (eds.): *Proceedings of the Fourth International Congress on Mathematical Education*, Birkhäuser, Boston, 247–249.
Niss, M. (1985): 'Applications and Modelling in the Mathematics Curriculum – State and Trends', *International Journal for Mathematical Education in Science and Technology* 18, 487–505.

Niss, M. (1987): 'Applications and Modelling in Mathematics Curriculum – State and Trends', *International Journal for Mathematical Education in Science and Technology* 18, 487–505.
Niss, M. (1989): 'Aims and Scope of Applications and Modelling in Mathematics Curricula', in Blum *et al.* (eds.), 22–31.
Niss, M , Blum, W. and Huntley, I. (eds.) (1991): *Teaching of Mathematical Modelling and Applications*, Ellis Horwood, Chichester.
Noss, R. *et al.* (eds.) (1990): *Political Dimensions of Mathematics Education: Action and Critique*, Institute of Education, University of London.
Orwell, G. (1987): *Nineteen Eighty-Four*, Penguin Books, London. (First edition 1949.)
Otte, M. (ed.) (1974): *Mathematiker über die Mathematik*, Springer, Berlin.
Paffrath, F. H. (ed.) (1987): *Kritische Theorie und Pädagogik der Gegenwart*, Deutscher Studien Verlag, Weinheim.
Peters, R. S. (1966): *Ethics and Education*, George Allan and Unwin, London.
Piaget, J. (1970): *Genetic Epistemology*, Columbia University Press, New York.
Plato (1992): *The Republic*, Introduced by Terence Irwin, Everyman's Library, London.
Plato (1992): *Theaetetus*, Edited with an Introduction by Bernard Willliams, Hackett Publishing Company, Indianapolis.
Pompeu, Jr., G. (1992): 'Bringing Ethnomathematics into the School Curriculum: An Investigation of Teachers' Attitudes and Pupils' Learning', Doctor Thesis, Department of Education, University of Cambridge.
Popper, K. R. (1963): *Conjectures and Refutations*, Routledge and Kegan Paul, London.
Popper, K. R. (1965): *The Logic of Scientific Discovery*, Harper and Row, New York. (First German edition 1934.)
Popper, K. R: (1972): *Objective Knowledge*, Oxford University Press, London.
Raith, W. (ed.) (1973): *Handbuch zum Unterricht: Modelle emanzipatorischer Praxis, Hauptschule*, Raith Verlag, Starnberg.
Restivo, S. *et al.* (eds.) (1993): *Math Worlds: Philosophical and Social Studies of Mathematics and Mathematics Education*, State University of New York Press, Albany.
Riess, F. (ed.) (1977): *Kritik des mathematisch naturwissenschaftlichen Unterrichts*, Päd. Extra Buchverlag, Frankfurt am Main.
Rousseau, J.-J. (1968): *The Social Contact*, Penguin Books, London. (First French edition 1762.)
Ryle, G. (1949): *The Concept of Mind*, Hutchinson, London.
Ryle, G. (1954): *Dilemmas*, Cambridge University Press, Cambridge.
Sapir, E. (1929): 'The Status of Linguistics as a Science', *Language* 5, 207–214.
Schumpeter, J. A. (1987): *Capitalism, Socialism and Democracy*, Unwin Paperbacks, London. (First edition 1943.)
Searle, J. (1969): *Speech Acts*, Cambridge University Press, Cambridge.
Searle, J. (1971): 'What Is Speech Act', in Searle, J. (ed.): *The Philosophy of Language*, Oxford University Press, Oxford, 39–53. (First published 1965.)
Searle, J. (1983): *Intentionality: An Essay in the Philosophy of Mind*, Cambridge University Press, Cambridge.
Shan, S.-J. and Bailey, P. (1991): *Multiple Factors: Classroom Mathematics for Equality and Justice*, Trentham Books, Stoke-on-Trent.
Skovsmose, O. (1980): *Forandringer i matematikundervisningen*, Gyldendal, Copenhagen.
Skovsmose, O. (1981a): *Matematikundervisning og kritisk pædagogik*, Gyldendal, Copenhagen.

Skovsmose, O. (1981b): *Alternativer og matematikundervisning*, Gyldendal, Copenhagen.
Skovsmose, O. (1984): *Kritik, undervisning og matematik*, Lærerforeningernes Materialeudvalg, Copenhagen.
Skovsmose, O. (1985): 'Mathematical Education versus Critical Education', *Educational Studies in Mathematics* 16, 337–354.
Skovsmose, O. (1988): 'Mathematics as Part of Technology', *Educational Studies in Mathematics* 19, 23–41.
Skovsmose, O. (1989a): 'Towards a Philosophy of an Applied Oriented Mathematical Education', in W. Blum *et al.* (eds.), 110–114.
Skovsmose, O. (1989b): 'Models and Reflective Knowledge', *Zentralblatt für Didaktik der Mathematik* 89(1), 3–8.
Skovsmose, O. (1990a): 'Mathematical Education and Democracy', *Educational Studies in Mathematics* 21, 109–128.
Skovsmose, O. (1990b): 'Reflective Knowledge: Its Relation to the Mathematical Modelling Process', *International Journal of Mathematical Education in Science and Technology* 21, 765–779.
Skovsmose, O. (1992): 'Democratic Competence and Reflective Knowing in Mathematics', *For the Learning of Mathematics* 2(2), 2–11.
Skovsmose, O (1993): 'The Dialogical Nature of Reflective Knowledge', in Restivo *et al.* (ed.), 162–181.
Snow, C. P. (1969): *The Two Cultures: And a Second Look*, Cambridge University Press, Cambridge. (First edition 1959.)
Sohn-Rehtel, A. (1970): *Geistige und körperliche Arbeit*, Suhrkamp, Frankfurt am Main.
Stork, H. (1977): *Einführung in die Philosophie der Technik*, Wissenschaftliche Buchgesellschaft, Darmstadt.
Swetz, F. J. (1987): *Capitalism and Arithmetic*, Open Court Publishing Company, La Salle, Illinois.
Taylor, F. W. (1947): *Scientific Management*, Harper and Brothers Publishers, New York. (Parts of the book first published in 1911.)
Thucydides (1972): *History of the Peloponnesian War*, Introduction and Notes by M. I. Finley, Penguin Books, London.
Tybl, R. and Walter, H. (eds.) (1973): *Handbuch zum Unterricht: Modelle emanzipatorischer Praxis*, Grundschule, Raith Verlag, Starnberg.
Tymoczko, T. (ed.) (1986): *New Directions in the Philosophy of Mathematics*, Birkhäuser, Boston.
Vithal, R. (1992): 'The Construct of Ethnomathematics, and Implications for Curriculum Thinking in South Africa', Master Thesis, Department of Education, University of Cambridge.
Volk, D. (1975): 'Plädoyer für einen problemorientierten Mathematikunterricht in emanzipatorischer Absicht', in Ewers, M. (ed.): *Naturwissenschaftliche Didaktik zwischen Kritik und Konstruktion*, Belz Verlag, Weinheim.
Volk, D. (1977): 'Entscheidungsraum in den Lehrplänen machen! Wissenschaftstheoretische Probleme der Matematik in der Ausbildung der Mathematiklehrer Aufnehmen', in Riess (ed.), 347–392.
Volk, D. (1979): *Handlungsorientierende Unterrichtslehre am Beispiel Mathematikunterrichts*, Band A, Päd Extra Buchverlag, Bensheim.
Volk, D. (ed.) (1979): *Kritische Stichwörter zum Mathematikunterricht*, Wilhelm Fink Verlag, München.

Volk, D. (1980): *Zur Wissenschaftstheorie der Mathematik: Handlungsorientierende Unterrichtslehre*, Band B, Päd Extra Buchverlag, Bensheim.
Volk, D. (1989): 'Mathematics Classes and Enlightenment', in Blum *et al.* (eds.), 187–191.
Volmink, J. (1989): 'Non-School alternatives in Mathematics Education', in Keitel *et al.* (eds.), 59–61.
Volmink, J. (1990): 'The Constructivist Foundation of Ethnomathematics', in Noss *et al.* (eds.), 243–247.
Volmink, J. *et al.* (1994): *Social, Cultural, and Political Issues in Mathematics Education: An Annotated Bibliography of Selected Writings, 1980–1990*, CASME, Durban.
Wagenschein, M. (1965, 1970): *Ursprüngliches Verstehen und exaktes Denken I–II*, Ernst Klett Verlag, Stuttgart.
Whorf, B. L. (1956): *Language, Thought and Reality*, Edited by J. B. Carrol, The MIT Press, New York.
Wilder, R. L. (1981): *Mathematics as a Cultural System*, Pergamon Press, Oxford.
Wittenberg, A. I. (1963): *Bildung und Mathematik*, Ernst Klett Verlag, Stuttgart.
Wittgenstein, L. (1961): *Tractatus Logico-Philosophicus*, Routledge and Kegan Paul, London. (First German edition 1921.)
Wittgenstein, L. (1953): *Philosophical Investigations*, Basil Blackwell, Oxford.
Young, M. F. D. (ed.) (1971): *Knowledge and Control*, Collier MacMillan, London.
Young R. (1989): *A Critical Theory of Education*, Havester Wheatsheaf, New York.

NAME INDEX

Abraham, J., 215, 227
Adorno, T. W., 11, 21, 28, 40–41, 76–77, 78, 97, 150, 192, 210, 227
Alrø, H., 224, 227
Andersen, B. W., 224
Andersen, J. J., 80
Angelis, D., 231
Anzinger, W., 211, 227
Apple, M. W., 227
Austin, J. L., 108, 167–169, 226, 227
Ayer, A. J., 227

Bailey, P., 215, 233
Bayle, P., 15, 206
Beck, U., 214, 227
Bell, D., 38–39, 50, 57, 213, 227
Benhabib, S., 210, 227
Benjamin, W., 214, 227
Benn, S. I., 212, 227
Berthelsen, J., 215, 227
Beth, E. W., 226, 227
Bibby, N., 215, 227
Bishop, A. J., 214, 216, 227, 228
Bjerre, O. L., 221, 228
Block, E., 211
Blomhøj, N., 220, 228
Bloor, D., 214, 228
Blum, W., 217, 228, 233
Boll, J., 143, 148
Boos-Bavnbek, B., 214, 228
Borba, M. C., 61, 215, 228
Bourdieu, P., 56–57, 214, 228
Bowles, S., 212, 228
Breugel, P., 59–60
Brousseau, G., 226, 228
Bruner, J. S., 74, 228

Bødtkjer, H., 125, 126, 127, 155

Carnap, R., 106–107, 167–169, 197, 218, 228, 229
Christiansen, B., 215, 223, 226, 229
Christiansen, I. M., 220, 226
Connerton, P., 210, 229
Copernicus, N., 42
Curry, H. B., 214, 229

D'Alembert, J. le R., 15
D'Ambrosio, U., 61, 189, 212, 215, 225, 229
Damerow, P., 215, 229
Darwin, C., 42
Davis, C., 229
Davis, P. J., 213, 214, 219, 229
Davis, Z., 231
Descartes, R., 197
Dessauer, F., 45, 214, 229
Dewey, J., 28, 36, 40, 211, 212, 229
Diderot, D., 15
Dilthey, W., 22
Dormolen, J., 228
Dowling, P., 216, 229
Dyhr, O., 80

Ellul, J., 45–46, 55, 229
Ernest, P., 216, 229
Euclid, 53
Evans, J., 216, 229

Fasheh, M., 210, 229
Fischer, R., 229
Foucault, M., 216
Frankenstein, M., 61, 215, 229, 230

Frege, G., 2
Freire, P., 22, 24, 26, 27, 60, 61, 77, 211, 230
Freudenthal, H., 215, 217, 230

Gadamer H.-G., 226, 230
Gehlen, A., 214, 230
Gerdes, P., 61, 215, 230
Giroux, H. A., 24–27, 28, 30, 41, 211, 212, 230
Glasersfeld. E. v., 224, 230
Glimm, J. G., 210, 230
Grouws, D. A., 230
Grønbæk Hansen, K., 224, 230

Habermas, J., 12, 21, 210, 211, 214, 230
Hanson, N. R., 103, 218, 230
Hardy, G. H., 42–43, 225, 230
Harris, M., 230
Hegel, G. W. F., 11, 15
Heidegger, M., 107
Held, D., 210, 212, 230
Hermann, K., 217, 230
Hersh, R., 213, 214, 229
Hilbert, D., 53, 54, 231
Hirst, P. H., 213, 219, 225, 230
Hoffman, M. R., 211, 215, 231
Hoffmann, D., 211, 231
Horkheimer, M., 11, 231
Howson, A. G., 229
Hume, D., 15, 202
Huntley, I., 217, 228, 233
Høines, M. J., 214, 231

Illeris, K., 22, 60, 211, 215, 227, 231

Jay, M., 210, 231
Jensen, H. S., 214, 231
Jensen, J. H., 211, 217, 231
Julie, C., 215, 231

Kant, I., 3–4, 11, 15, 198, 206, 231
Kapp, E., 45, 46, 231
Keitel, C., 214, 217, 231

Kellner, D., 210, 231
Kitcher, P., 219, 231
Klöcker, M., 60, 62
Kotzmann, E., 214, 217, 231
Kristoffersen, I. L., 60, 62
Krogshede Nielsen, A. M., 224
Kuhn, T. S., 231

Lakatos, I., 51, 111, 218, 231
Lange, J. de, 217, 232
Lave, J., 219, 232
Leibniz, G. W., 216
Lempert, W., 21, 25, 60, 211, 232
Lerman, S., 216, 232
Lindenskov, L., 182, 224, 232

Malcolm, N., 218, 232
Malle, G., 229
Marcuse, H., 11, 45–46, 232
Marx, K., 11, 15, 17, 18, 22, 33, 47, 211
McPeck, J. E., 217, 232
Mellin-Olsen, S., 214, 215, 223, 224, 225, 228, 231, 232
Mill, J. S., 35, 212, 213, 232
Mills, C. W., 76, 232
Mollenhauer, K., 21, 60, 211, 232
Moore, G., 13–14, 22, 210, 232
Münzinger, W., 215, 232

Negt, O., 21, 60, 76–78, 82, 211, 232
Newton, I., 18
Nickson, M., 216, 224, 232
Nietzsche, F., 3
Niss, M., 57, 214, 215, 216, 217, 228, 230, 232, 233
Noss, R., 215, 233

Orwell, G., 1, 5, 12, 16, 209, 210, 233
Otte, M., 214, 226, 228, 229, 233

Paffrath, F. H., 210, 233
Pate, G., 214, 228
Pericles, 31, 32

NAME INDEX

Peters, R. S., 212, 227, 233
Piaget, J., 62, 183, 203–205, 216, 224, 226, 227, 233
Plato, 13, 22, 31, 101, 197, 233
Pompeu, Jr., G., 61, 233
Popper, K. R., 51, 52, 55, 112, 214, 233
Poulsen, S. C., 215, 227
Powell, A. B., 61, 215, 230, 231

Raith, W., 211, 233
Rauch, E., 211, 227
Reinholt, A., 80, 155
Restivo, S., 214, 233, 234
Riess, F., 215, 233, 234
Rousseau, J.-J., 31–32, 150, 212, 233
Russell, B., 2, 101, 106, 167–169, 218
Ryle, G., 108, 167–169, 223, 233

Sapir, E., 2–3, 210, 233
Sartre, J. P., 223
Schumpeter, J. A., 36–37, 38, 39, 213, 233
Searle, J., 223, 226, 233
Shan, S.-J., 215, 233
Skovsmose, O., 214, 215, 217, 224, 227, 231, 233, 234
Smith, A., 13
Snow, C. P., 6, 234
Sohn-Rehtel, A., 214, 234
Spinoza, B. de, 7

Stork, H., 45, 214, 234
Swetz, F. J., 218, 234

Taylor, F. W., 47, 234
Thucydides, 31, 212, 234
Trankjær, I., 224
Tybl, R., 211, 234
Tymoczko, T., 214, 234

Vico, G., 48
Vithal, R., 61, 215, 234
Vognsen, J., 143, 148
Volk, D., 215, 234, 235
Volmink, J., 61, 215, 234
Voltaire, F. de, 15

Wagenschein, M., 22, 75–78, 211, 216, 224, 235
Walter, H., 211, 234
Walther, G., 223, 226, 229
Werner, T., 215
Whorf, B. L., 2–3, 210, 235
Wilder, R. L., 214, 235
Wittenberg, A. I., 216, 235
Wittgenstein, L., 2, 9–10, 106–108, 169, 179, 218, 225, 226, 235

Young, M. F. D., 235
Young R., 211, 235

Ørberg, T., 80, 216

SUBJECT INDEX

Absolutism, 19–20, 82, 101, 199, 226
abstraction, 50–56
   real, 54, 218
   realised, 50–56, 95, 102, 105, 108, 136, 214, 218
   thinking, 50–56, 95, 105, 108, 136, 214, 218
achievement, 189–191
action, 176–184, 222–223, 224
   and intention, 176–181
   and disposition, 176–181
   and learning, 181–184, 190, 192
activity, 176, 178–179, 183, 184–187, 223–224
   blind, 186
   directed, 185–187
   forced, 185, 189, 201, 208
   prescribed, 185
algorithm, 104, 109, 113, 115, 118–122, 127–128, 131–132, 134–136, 138, 139, 152
algorithmic language, 109–111, 139
Allgemeinbildung, 40, 74, 213

Background, 179–180, 190–191
basis for critique and correction, 111, 113–114
blind activity, 186
broken intention, 187, 190

Capitalism, 23, 33–34, 77, 102
challenging questions, 138–140
concreteness, 62–63
   physical, 62–63
   social, 62–63
constructivism, 118, 204, 215–216

crisis, 11-27, 77, 209, 210, 217
   dialectics of crises, 20–21
   hierarchical order of crises, 17–19, 20, 22–23
critical activity, 14–19, 21, 27, 76, 208–209
critical citizenship, 24, 36, 37, 38, 40, 41, 59, 74, 78, 97, 101, 114, 175, 192, 219, 222
critical key-terms, 61
   critical competence, 40, 61, 154, 190
   critical distance, 61
   critical engagement, 61
critical situation, 14, 17, 18, 20, 22, 24, 44, 172, 175
critical theory, 17
Critical Theory, 11–12, 21, 22, 76, 150, 215
critique, 11–27, 28, 41, 44, 56, 58, 59, 100, 141, 151, 172–173, 175, 183, 192, 196, 198, 201, 205–206, 208–209
   and democracy, 28, 41, 58, 59
   and education, 11–27
   object of, 44, 56, 113, 175, 207, 208
   once-and-for-all, 198, 200–201, 206
   subject of, 175, 192, 201, 208
critique of ideology, 17, 19, 22, 25, 27, 141, 150, 212

Data, 133–134
democracy, 21, 25, 27, 28–41, 57–58, 59, 114, 141–142, 148–

152, 212, 213, 222
and capitalism, 33–34
and critique, 28, 41, 58, 59
and delegation of sovereignty, 32, 34, 37, 149
and education, 21, 25, 27, 28–41, 148–152
and expertocracy, 38–39, 99, 222
basic ideas of, 31–34
in a highly technological society, 33, 37, 38–40, 57–58, 99, 114, 213
democracy, conditions for, 28–29, 33–34, 36–37, 38, 39
ethical, 29, 38
formal, 29, 33, 36, 38
material, 29, 33, 38
possibility for participation, 29, 38
democracy, types of
critical, 31
direct, 32, 36, 37, 142
radical, 25, 27
representative, 32–34
democratic competence, 30–31, 34–37, 40–41, 57–58, 74, 78, 114, 118, 122, 141, 149–150, 151, 213
democratic life, 29, 30–31, 33, 34, 35–36, 37, 39, 40, 142, 149
determinism, 18–19, 44, 48, 100, 137
dialogue, 205–206
directed activity, 185–187
dia-logical epistemic theory, 205–206, 207
disposition, 176–181, 182, 184, 190–191, 223, 224
background, 179–180, 190–191
foreground, 179–180, 190–191, 193–194

Education
and critique, 11–27
and democracy, 21, 25, 27, 28–41, 148–152
equality in, 30
emancipation, 12, 19–21, 25–26, 45, 211
empiricism, 42, 196–198, 202–204
energy technology, 46–47, 49
Enlightenment, 15, 206
entry points to reflective knowing, 118–122
epistemic development, 176, 184–189, 190
epistemic theory
dia-logical, 205–206, 207
mono-logical, 201–205, 206
equal opportunity, 29–30
equality in education, 30
ethnomathematics, 56, 61–62, 189
exemplarity, 21, 22, 73–78, 82, 94, 97, 155, 172, 173
expertocracy, 38–39, 99, 222

Fachkritik, 21, 211
fallibilism, 82, 101, 199
fatalism, personal, 69, 189–190
fata morgana, 168–169, 170
forced activity, 185, 189, 201, 208
foreground, 179–180, 190–191, 193–194
formal language, 2, 53–55, 102, 106–111, 135, 137–138, 166–170, 171, 221
formalism, 42, 53, 110
formalisation, 53–56, 57, 105, 106–108, 119, 171
of actions and routines, 54, 105, 222
of language, 53–55, 105, 106–108, 222
of mathematics, 53–55
formatting
global, 102
local, 102
formatting power,
of mathematics, 42–58, 59, 78, 82, 96, 105, 117–118, 120–

SUBJECT INDEX

122, 125, 128, 135, 136–138, 141–142, 148, 150–151, 152, 166–170, 175, 207, 213, 214
of mathematics education, 190

Genetic epistemology, 62, 183–184, 203–205, 224

Highly technological society, 5–6, 23, 33, 37, 38–40, 41, 50, 52, 56–58, 98, 99, 114, 213, 222
homogeneity of knowledge, 196–197, 200–201, 205
hypothetical situation, 170–171, 222

Ideology, 3, 16, 26, 27, 211
critique of, 17, 19, 22, 25, 27, 141, 150, 212
ignored intention, 187
information society, 38–39, 57
information technology, 43, 46–47, 49–50, 55, 100
instrumental intention, 188–189
instrumentalism, 188–189, 190, 225
integrated intention, 187
intention, 176–189, 201, 223
broken, 187, 190
ignored, 187
integrated, 187
instrumental, 188–189
modified, 187, 190
shared, 187, 193, 194
underground, 188, 190, 225
intention of learning, 181–184, 193
intention in learning, 183–184, 193
intentional explanation, 180–181
intentionality, 175–195, 208, 221, 223
interdisciplinarity, 17, 60, 76, 79, 142, 151
inverse alchemy, 135, 167

Knowing, 122–124, 196–209
a web of interrelationship, 123
an explosive concept, 206–209
an open concept, 199–201
knowing, politics of, 205, 209, 216
knowing, mathematical, 102, 114–116, 123–124, 125, 164–165, 196, 217, 219
knowing, reflective, 97–124, 125, 133–136, 140, 141–142, 149–151, 152–154, 164–165, 170, 171, 173–174, 175, 192, 196, 201, 208, 218, 219, 226
knowing technological, 102, 114–116, 123–124, 125, 164–165, 196, 219
knowing, types of, 101–102, 122–124, 125, 196
knowledge, 196–209, 225, 226
a classical definition of, 197–198, 199, 200
a controlled concept, 196–198
an authorised body of, 198, 199, 200–201, 225
homogeneity of, 196–198, 200–201, 205
knowledge, types of, 98–101
knowledge, mathematical, 100–101, 203–205, 217, 219, 226
knowledge, reflective, 99–101, 217
knowledge, technological, 57, 98–101, 217
knowledge conflict, 200, 205–206

Language
algorithmic, 109–111, 139
formal, 2, 53–55, 102, 106–111, 135, 137–138, 166–170, 171, 221
formalisation of, 53–55, 105, 106–108, 222
mathematical, 4, 106, 109–111, 113, 139
natural, 106–111, 113, 135, 137–138, 166–170, 171, 221
systemic, 109–111, 134, 218,

220
language games, 53, 106–111, 114, 125, 133–136, 139, 166, 199
   and modelling, 106–111, 114, 125, 133–136, 139, 166
   transition between, 109–111, 114, 125, 133–136, 139, 166
learning as action, 181–184, 190, 192
liberation
   and technology, 44–46
linguistic relativism, 2–5, 43, 48, 56, 108, 137, 207, 217
literacy, 24–27, 41, 59, 60, 97, 117, 122, 137, 172, 192, 207, 212, 220
logical positivism, 2, 169, 180, 210, 217

Mathemacy, 24–27, 41, 59, 60, 74, 97, 98, 114, 117, 118, 122, 124, 125, 137, 138, 141–142, 150, 154, 170–174, 175, 192, 207–208, 212, 219, 220, 221, 222
   and reflective knowing, 117, 124, 125, 141–142, 150, 154, 171, 173, 192
mathematical archaeology, 94–96, 167, 194, 216, 217
mathematical knowing, 102, 114–116, 123–124, 125, 164–165, 196, 217, 219
mathematical knowledge, 100–101, 203–205, 217, 219, 226
mathematical language, 4, 106, 109–111, 113, 139
mathematical modelling, 51, 52, 102–114, 133–136, 141, 154, 167, 207, 217
   algorithm, 104, 109, 113, 115, 118–122, 127–128, 131–132, 134–136, 138, 139, 152

algorithmatisation, 104
   and reflective knowing, 102–114, 118–122, 133–136
   basis for critique and corrections, 111, 113–114
   interpretation, 104, 109
   language games in, 106–111, 114, 125, 133–136, 139, 166
   mathematical model, 55, 104, 106, 108, 111, 112
   mathematisation, 63, 70, 104, 109, 110, 134
   problem area, 102–103, 109, 113, 173
   problem identification, 104, 111–112, 218
   realisation, 104
   scope of possible actions, 111, 113
   structure of argumentation, 111–112
   system, 103, 109
   system development, 103–105, 106, 109, 110, 120, 134–136, 138, 218, 220
mathematical modelling, types of
   extended, 102, 105, 218, 219
   pointed, 102, 105, 111, 113, 120, 218, 219
mathematics,
   and modelling, 51, 52, 102–114, 133–136, 141, 154, 167, 207, 217
   and technology, 4–5, 48–50, 78, 98–102
   as a formatting power, 42–58, 59, 78, 82, 96, 105, 117–118, 120–122, 125, 128, 135, 136–138, 141–142, 148, 150–151, 152, 166–170, 175, 207, 213, 214
   as critical, 56–58
   children's meta-conceptions of, 80–82
   formalisation of, 53–55

mathematisation, 63, 70, 104, 109, 110, 134
meaning
  and scene-setting, 91–93, 149
modified intention, 187, 190
mono-logical epistemic theory, 201–205, 206
multi-faceted semantics, 94, 123, 154, 216
Mündigkeit, 40–41, 58, 78, 150, 192, 208

Natural language, 106–111, 113, 135, 137–138, 166–170, 171, 221
negotiation, 205–206

Object of critique, 44, 56, 113, 175, 192, 207, 208

Personal fatalism, 69, 189–191
philosophy
  and educational practice, 7–10, 59–60, 97–98
  analytical, 8–9
  of mathematics education, 6–10, 59–60
  of technology, 43–48
picture theory, 9, 108, 167
Platonism, 42, 55
politics of knowing, 205, 209, 216
prescribed activity, 185
principle of hope, 20, 211
problem orientation, 21, 22, 78, 92, 142, 218
project organisation, 21, 22, 78, 92, 142
project work, 74, 78, 79, 97, 166, 214–215, 216
power, symbolic, 56–58, 59
public stratification, 68–69

Rationalism, 7, 196–198, 202–204
real abstraction, 54, 218
realised abstraction, 50–56, 95, 102, 105, 108, 136, 214, 218

reflective knowing, 97–124, 125, 133–136, 140, 141–142, 149–151, 152–154, 164–165, 170, 171, 173–174, 175, 192, 196, 201, 208, 218, 219, 226
  and mathematicy, 117, 124, 125, 141–142, 150, 154, 171, 173, 192
  and modelling, 102–114, 118–122, 133–136
  and challenging questions, 138–140
  entry points to, 118–122
reflective knowledge, 99–101, 217
relativism, linguistic, 2–5, 43, 48, 56, 108, 137, 207, 217
risk society, 20, 214, 222
risk structure, 20, 50, 56, 98, 100, 222

Scene-setting, 91–94, 115, 117, 123, 125, 133, 137–138, 149, 152, 154, 166, 175, 193–195, 216
  and meaning, 91–93, 149
  and vantage point, 166, 193
scene-setting, types of, 92, 149
  activity based, 92, 149
  imaginative, 92, 149
  realistic, 92, 149
semantics, multi-faceted, 94, 123, 154, 216
servility, 189–191
shared intention, 187, 193, 194
social technology, 46–47, 49
society
  highly technological, 5–6, 23, 33, 37, 38–40, 41, 50, 52, 56–58, 98, 99, 213, 222
  information, 38–39, 57
sociological imagination, 76, 77–78, 172–173, 208
stratification, public, 68–69
structuralism, 74–75, 115, 116
student's curriculum, 182

subject of critique, 175, 192, 201, 208
symbolic power, 56–58, 59
system development, 103–105, 106, 109, 110, 120, 134–136, 138, 218, 220
systemic language, 109–111, 134, 218, 220

Technological absorption, 138–139
technological determinism, 44, 48, 100, 137
technological knowing, 102, 114–116, 123–124, 125, 164–165, 196, 219
technological knowledge, 57, 98–101, 217
technological optimism, 6, 18, 44, 48, 100, 208
technological pessimism, 19, 44, 46, 48, 50, 208
technology,
and liberation, 44–46
and mathematics, 4–5, 48–50, 78, 98–102
philosophy of, 43–48
technology, types of, 46–47
energy, 46–47, 49
information, 43, 46–47, 49–50, 55, 100
social, 46–47, 49
tools, 45, 46, 47, 48–49
thematic approach, 59–78, 79, 97, 216
thinking abstractions, 50–56, 95, 105, 108, 136, 214, 218
tools, 45, 46, 47, 48–49

Underground intention, 188, 190, 225

Vantage point, 165–166, 173, 175, 193
Vico paradox, 43–48, 50, 58, 98, 100, 102, 208–209, 214